Naked Safety

Workplace safety has never been seen as sexy, clever or cool.

Fraught with legislative hurdles, ambiguous policy and complex procedures, despite its alleged importance safety has lost its way. For many organisations safety is seen as burdensome and bureaucratic and has become little more than paperwork and performance charts: things done in fear of persecution – from the authorities, the media or the civil arena – rather than doing the right thing.

To change the game and build real risk literacy, it's vital to make things easier, to strip things back to basics and think again about how we work. This is *Naked Safety*.

Encouraging the reader to step outside their comfort zone, this book demystifies workplace safety, challenging traditional views and catalysing critical thought and high-impact action.

With narratives on the central pillars of workplace safety including risk management; legal frameworks; performance; governance; leadership and culture, as well as perspectives on key issues that affect safety – and business – more broadly, such as worker wellbeing; employee engagement; the impact of globalisation; corporate social responsibility; sustainability and the role of the safety practitioner, *Naked Safety* features over 100 actions to bring about positive, sustainable organisational change.

This book is a useful, multi-purpose guide for professionals and an indispensable toolkit for practitioners, business leaders and anyone with an interest in workplace risk and Occupational Safety and Health. Let's get Naked!

Andrew Sharman teaches leadership and safety culture at the European Centre for Executive Development (CEDEP) in Fontainebleau, France, and consults to Fortune 500 corporations and non-governmental organisations around the world. He is the best-selling author of several books on leadership and safety culture.

Naked Safety

Exploring the Dynamics of Safety in a Fast-Changing World

Andrew Sharman

LONDON AND NEW YORK

First published 2019
by Routledge
2 Park Square, Milton Park, Abingdon, Oxon OX14 4RN

and by Routledge
711 Third Avenue, New York, NY 10017

Routledge is an imprint of the Taylor & Francis Group, an informa business

British Library Cataloguing-in-Publication Data
A catalogue record for this book is available from the British Library

Library of Congress Cataloging-in-Publication Data
Names: Sharman, Andrew, author.
Title: Naked safety : exploring the dynamics of safety in a fast-changing
world /Andrew Sharman.
Description: Abingdon, Oxon ; New York, NY : Routledge, 2018. |
Includes bibliographical references and index.
Identifiers: LCCN 2018008794 | ISBN 9780415827775 (hardback) |
ISBN 9781315858692 (ebook)
Subjects: LCSH: Industrial safety. | Corporate culture.
Classification: LCC T55 .S44825 2018 | DDC 658.3/8—dc23
LC record available at https://lccn.loc.gov/2018008794

ISBN: 978-0-415-82777-5 (hbk)
ISBN: 978-1-315-85869-2 (ebk)

Typeset in Times New Roman
by Florence Production Ltd, Stoodleigh, UK

To those who seek to strip things back and engage others in creating safety

Contents

About the author

Andrew Sharman is an in-demand consultant, speaker and coach with more than 20 years' experience in occupational safety in more than 100 countries. He's seen first-hand the overwhelming, stultifying approach that many organisations get trapped into taking with safety so he works with leaders and teams in NGOs and Fortune 500 corporations around the globe – including the coolest technology brand, the fastest Formula One team and market leaders in the energy, pharmaceutical, FMCG and mining sectors – to change the game, helping them strip back their approach to safety and risk management to improve their culture and enable excellence.

Andrew is Professor of Leadership and Safety Culture at the European Centre for Executive Development (CEDEP) on the INSEAD campus in Fontainebleau, France. He is Chairman of the Board of the Institute of Leadership & Management and a Chartered Fellow of the Institution of Occupational Safety & Health, for which he also serves as Vice President.

Far from being risk-averse, Andrew loves adventure sports including free flying, sea kayaking and swimming with sharks. Contact him at www.andrewsharman.com.

Preface

"And te tide and te time þat tu iboren were, schal beon iblescet."

St Marher, 1225

This book no longer belongs to me. It's yours. So, read it as you wish. Underline points of interest, scratch out what gets your goat, write in the margins. Think whatever you want to think. All I ask is that you *do something* as a result of what you read in these pages.

This book is intended to be a practical tool, rather than a theoretical one. The aim has been to compile the most relevant research, share actionable insights and provide a practical framework designed to increase the chance of success in creating workplaces that engage, encourage, empower and enable people to work well, feel happy and return home without harm at the end of every shift.

Naked Safety is not a leadership theory or safety model, it's a way of thinking and working. It's an antidote to the oppression and frustration experienced by shop-floor employees, supervisors, managers and senior leaders in organisations around the world when faced with the burdens of bureaucracy that have been created in pursuit of eliminating workplace injuries and ill-health. It's not about laws, rules and procedures. It's not about 'making people safe'. It's not about preventing accidents, it's about *creating safety*.

Like my other books, I hope that this book is a primer for thought. But it is not written to be thought-provoking. It is written to be action-provoking. And in support of this, you'll find a series of actions at the end of each chapter – there are over 100 in total. These actions don't come with an iron-clad guarantee, however: I'm not a guru, savant or academic, I'm simply someone who's had the privilege to work with hundreds of organizations and thousands of safety practitioners and business leaders around the world and discover what works and what doesn't. I've endeavoured to share these ideas, together with the *where, when, how* and *why* in this book. You'll have to enter into the spirit of things and adjust the actions to suit yourself, your organisation and your needs.

It's time now to strip things back, to show that workplace safety is a significant enabler to business success, a route to boosting worker engagement, job satisfaction, morale and general wellbeing. And it's something in which *you* have an important role to play. As Saint Marher warns, and as Mark Twain reminds, '*time and tide wait for no man*', so come on, let's get *naked*.

Acknowledgements

This book was formed with the idea of pushing back against the tide of negativity surrounding workplace safety by sharing and developing ideas from some of the greatest thinkers in the domain of leadership, work psychology and risk management. To this end I am especially indebted to the work of John Adair, Åsa Boholm, Rhona Flin, Geert Hofstede, Erik Hollnagel, Gary Klein, Kurt Lewin, Jens Rasmussen, James Reason, Burrhus Skinner, Paul Slovic and Karl Weick.

Over the years I've built a bank of gratitude to certain people who have continued to push, provoke and support me with their words and deeds. Many thanks go to Emer Callaghan, Dom Cooper, Sidney Dekker, Frank Furedi, Dame Judith Hackitt, Leandro Herrero, Pat Hollingworth, Andrew Hopkins, Judi Komaki, Evelyn Kortum, Alan Quilley and Ed Schein for straight-talking and stimulating discussions. Full marks go to Steve Giblin for sparking an idea and continuing to fan the flames.

Finally, to the business leaders and safety practitioners around the world with whom I have worked over the last 20 years: as you have shared your perspectives and ideas, your fingerprints have covered the pages of this book. Thanks for getting *naked* with me.

Part 1

Stripping back

1 Introduction

Recent research by the International Labor Organization (ILO) – a specialised United Nations agency comprising of 187 member countries – suggests that there are around 2.3 million deaths around the world each year as a result of occupational accidents or diseases relating to the workplace.

Assuming the ILO's figures are correct, that would mean that every 15 seconds 153 workers have a work-related accident – and, within in that same 15 seconds, one will result in a fatality.

However, in most developing countries health and safety reporting is limited, and, as a result, it's difficult to garner accurate details about accidents, deaths and illnesses – so, the likelihood is that this figure is actually a significant underestimate.

Is 'good safety good business'?

Beyond the obvious human tragedy, there is also a huge economic cost. The ILO estimate that the average cost of workplace accidents is around 4 per cent of the global Gross Domestic Product (GDP) – an estimated 2.8 trillion US Dollars.

In Great Britain, the workplace regulator, the Health & Safety Executive (HSE), calculates that the annual cost of workplace accidents, injuries and work-related ill-health is around £13.8 billion.

With statistics like these, it is surprising that the idea that 'good safety is good business' is slow to resonate.

An extensive study by The World Economic Forum recently discovered that there is a definite link between 'organisational competitiveness and workplace health and safety' – that low accident rates make for higher productivity.

The study concludes by stating that, within organisations, safety should be seen as a 'strategic asset' – and recognised as central to fostering competitiveness and innovative capacities.

As well as demonstrably increasing productivity, competency levels and commitment in workers within an organisation, good safety also enhances reputation and adds brand value.

If this wasn't reason enough for organisations to adopt more wide-reaching safety measures, the incentives for not doing so are also slowly eroding – with regulators in successive states around the world now beginning to clamp down on poor safety standards.

Though, in theory, penalties have been in place to stop reckless managers and business leaders from eschewing safety practices within organisations for many years – traditionally, bad practice and negligence has rarely resulted in meaningful legal action for those at the top.

However, in many countries there has been something of a sea change of late. In Britain, for example, in the 12-month period from April 2015 to March 2016, 46 business leaders and company directors were prosecuted by the HSE – of which, 12 were summarily handed prison sentences.

Fines for safety offences have also increased by around 43 per cent over the same period – and, in Britain, are now regularly hitting the £1 million mark, even for non-fatal accidents.

Relocation, relocation, relocation

However, tightening of regulations and stricter enforcement has had a further side-effect. Studies by the World Health Organization (WHO, 1999) indicate that to reduce costs – and, perhaps, to side-step potential legal complications – many large organisations are simply relocating their operations to developing countries – which are now home to a staggering 75 per cent of the global workforce.

Without a doubt, deaths and injuries already take a particularly heavy toll in developing countries, where large numbers of workers are concentrated in primary and extractive activities, such as agriculture, fishing, logging and mining – the most hazardous industries.

The ILO reports: "Agriculture is one of the most hazardous occupations worldwide. An estimated 1.3 billion workers are engaged in agricultural production worldwide – half of the total world labour force."

The ongoing relocation of big business has meant that the global workforce is in a state of flux. As barriers to trade and movement of products and supplies have fallen, this has increased the opportunities for transience as workers cross borders in search of work, causing an exponential rise in health, safety and security challenges.

As the proportion of migrant workers within the labour force continues to grow, many workplaces are becoming populated by individuals from many diverse backgrounds – which increases propensity for accidents due to cultural misunderstandings and poor communication. For example, recent research (Horck, 2006) reveals that misunderstandings owing to cultural differences play a significant role in approximately 80 per cent of maritime accidents.

Cultural differences have also lead to an increased Occupational Safety and Health (OSH) risk, due to the fact that in multicultural work teams, individual perceptions of what is 'safe' often radically differ.

Due to the changes in the nature of the workplace, socio-economics and regulatory systems, plus demographic developments and labour market shifts, the risk-based safety approach – advocated by the likes of Beck, Dekker, Reason and also in my own previous books – that has historically been the cornerstone of the world's OSH management is now being reconsidered.

The uncertain present and unpredictable future

Previously unknown levels of uncertainty and unpredictability haunt the global economy – and this has negatively impacted OSH throughout the world. Most organisations now face a dilemma over how best to maintain first-rate safety systems whilst endeavouring to reduce spiralling costs.

The ongoing global economic crisis has added to these challenges, simultaneously placing pressure on workers, families and organisations and undermining established patterns of behaviour, prevention systems, programmes and practices.

In such challenging times, safety is often side-lined and reduced to little more than an 'add-on' activity – even despite the fact that most academic and organisational research makes it clear that safety, when pursued proactively, has the power to drive and deliver competitive advantage – and reduce operational costs.

The move to make 'good safety' synonymous with 'good business' had made some progress following the publication of the high-profile Cullen Reports,[1] but in an uncertain climate, many organisations feel as though they are caught in a perfect storm – with barrage of potential hazards coming from multiple directions.

In addition to potential financial and reputational worries, in recent years, we have seen epidemics (such as the SARS virus), pandemics (like swine and avian 'flu), natural disasters (such as earthquakes, volcanic eruptions and tsunamis), man-made disasters (from oil spills to factory collapses), product recalls (from everything from cars to children's toys to baby milk) and workplace accidents – all of which have the potential to bring organisations to a shuddering halt.

Though all the above events may inversely affect organisations, they have all helped ferment the idea of organisational safety, contributed to the dialogue and allowed it to permeate culture.

However, even despite this, many organisations have markedly different ideas about what 'safety' actually is. Perhaps this is unsurprising. In a marketplace dominated by multinational corporations, is it possible to agree on a definition? Considering the variety of factors involved – including geography, workplace demographics, local culture, values, beliefs and supply-chain infrastructures – how could one realistically forge a single 'catch-all' safety standard?

Societal fear and the erosion of real safety

There has also been a pronounced backlash against the safety industry recently. Anger directed at what is pejoratively termed 'the nanny state' has been luridly illustrated with examples of 'safety' overcomplicating things and getting in the way of constructive or innocuous activities.

Popular examples of such things as banning yo-yos from schools, outlawing hanging baskets and removing knives from restaurant kitchens, have fed a common outrage concerning lost liberties and caused much derisive talk of 'health and safety gone mad'.

With these sorts of stories making regular newspaper headlines, it's little surprise that health and safety has become a major concern for many people and organisations. More now than ever before, there is an overriding fear that safety regulation has the power to interfere with anything it turns its sights on – and, if it suits its purposes, stop it in its tracks.

In *Culture of Fear: Risk-taking and the Morality of Low Expectation* (2007), sociologist Frank Furedi states that: "fear has become an ever-expanding part of life in the twenty-first century". I think he's dead right.

Furedi believes that, despite enjoying unprecedented levels of safety and security, modern society is subjugated by an all-encompassing narrative of fear, which has led to a culture that routinely inflates the dangers and risks of the general population. He argues that fear shifts easily from one problem to the next

> without there being a necessity for causal or logical connection. It becomes a sort of transitory, self-fulfilling aspect of modern life that is constantly sustained by a culture that is anxious about change and uncertainty – and continually chooses to anticipate the worst possible outcome.

As a consequence, hesitation and caution are seen as virtues and any risk-taking behaviour looks like reckless and irresponsible behaviour.

Under the auspices of 'risk management', fear becomes institutionalised and the fear response becomes further pronounced – transforming into 'risk aversion'. The need to eliminate risk starts to fixate business leaders and safety teams until it entirely dominates their collective psyches.

However, there is a real danger here. That danger is that people, society and organisations will begin to turn away from occupational safety and risk management due to misrepresentative media reporting.

The depiction of health and safety as an intrusive and disruptive waste of time means that safety ideals are often misunderstood – and, indeed, often maligned. Consequently, with focus on growing development, neoliberal policy agendas around the world have encouraged the removal of the 'burdens of bureaucracy' and other organisational restrictions. In some jurisdictions, the monitoring of safety standards has been drastically cut back or – more worryingly – replaced by 'voluntary' industry compliance.

But we cannot blame media spin and social misperceptions for everything. I can't help but feel that there has also been a push in the wrong direction from *within* organisations too. Has the lack of understanding of what safety really is about from operational managers and leaders actually encouraged those charged with building efficient health and safety management systems and improving performance to shroud what they do in mystery – effectively turning the discipline and activities into a black art? As this 'stealth and safety' takes hold, more managers turn their backs on getting involved, leaving things to the 'technical expert' and instead focusing on what they know best – their own domain or area of responsibility.

Encouraging a different perspective

To truly understand the issues around safety, it is important to look at the broader organisational culture, the current operating climate, the influence of global geographies, communities and psychologies.

It is also essential to engage with people and to keep seeking first-hand knowledge. After all, it is only when safety practitioners have developed 'risk curiosity' that they stand any chance of attaining 'risk literacy'.

Despite some popular opposition in the media, safety is not broken. It is not a case of throwing it all out and starting again. Indeed, some incredible progress has been achieved in recent years – with many organisations, industries and countries continuing to build, develop and share knowledge. The focus on safety is continuing to sharpen.

The purpose of this book is not to set out a spirited defence of organisational health and safety. Neither does it exist to instruct the reader on how to build a safety management system. Nor is it a list of regulatory requirements, or a 'how-to' guide on doing risk assessment or accident investigation.

This book is fundamentally concerned with encouraging people to consider risks from different perspectives – and to promote discussion and 'positive disruption' within the field of workplace Occupational Safety and Health.

It seeks to challenge the perceived wisdom on safety, spark discussion and debate – and shine a light on new ways to consider the issues around organisational safety.

The challenge for anyone working in the safety sphere today is always to find ways, not only to dispel the myths, but also confront the apathy that often grows up around the subject. Both of which often result in inertia.

If progress is to be made, safety professionals need to operate in a more forward-thinking and inclusive way, with a view to building working environments where safety is viewed as an immutable condition of the successful completion of tasks – rather than a potential bonus or an orbiting added-extra.

Too many safety practitioners spend their time debating the merits of different approaches and systems, instead of focusing on the most appropriate strategy based on an individual organisation's maturity, industry, function and culture.

Creating safety and getting naked

In my previous book *From Accidents to Zero* I introduced the concept of *creating safety*. This notion repositions safety as an inputs-focused activity rather than a dogmatic fascination with the avoidance of adverse events and the reduction of accident rates.

To truly create safety, and to build real risk literacy, it is vital to make things easier, to strip what we do back to founding principles, establish an 'open source' mentality, keep safety systems simple and appropriate – and invite contributions from everyone. This is what is meant by *Naked Safety*.

So, this book will strip back the key topics related to workplace safety, providing you with a range of different perspectives to consider. You'll find narratives on the central pillars of health and safety, including risk management; legal frameworks; performance; governance; leadership; and culture, as well as perspectives on key issues that affect health and safety – and business – more broadly. These chapters encompass elements such as worker wellbeing; employee engagement; the impact of globalisation; corporate social responsibility; sustainability; and the role of the health and safety practitioner.

The chapters conclude in summary, and then provide the reader with a series of *Naked Truths*, the real keys to the narrative. And then, to help you strip things back in your own organisation you'll *get naked* with a set of practical points to review in your workplace.

Are you ready to start? Come on, let's get *naked*.

Note

1 The reports of public inquiries into the Piper Alpha, Dunblane and Ladbroke Grove disasters overseen by former Lord Justice General, William Cullen, Baron Cullen of Whitekirk.

2 Globalisation

Over the last few decades, globalisation has been responsible for a trend amongst western organisations to relocate parts of their operation beyond domestic markets to other regions around the globe.

Typically, this involves shifting production into labour-intensive sectors in the developing world, where the cost of transporting goods is offset against the low wages of the local workers – and where enforced safety legislation is often negligible.

In this chapter, we'll consider how globalisation has affected – and continues to affect – the global economic landscape, how it has impacted both rich and poor regions and what it is likely to mean for the future of Occupational Safety and Health, if greater regulation isn't imposed on the operations of multinational organisations setting up abroad.

The International Labor Office (ILO) calculates that around 2.2 million people die of work-related accidents and diseases each year – that's approximately 5,000 people every day.

Depending on how fast you read, around 100 people will lose their lives due to work related-accidents and diseases just whilst you are reading this chapter.

As shocking as these numbers are, due to poor reporting in many regions, that number is almost certainly a huge underestimate. This poor focus on reporting has only been exacerbated in recent decades by the rapid development and strong competitive pressures of globalisation.[1]

Both the ILO and the World Health Organization (WHO), United Nations agencies with special interest in the areas of work and health, have warned that the pressures for deregulation of basic standards for health and safety is growing, and that globalisation will inevitably add to an upsurge in the number of work-related diseases, injuries and fatalities in the future. I suspect that they're right.

The politics of power and uneven playing fields

In the pursuit of commercial advantage, manufacturing in the new global economy has shifted from regulated, well-paying, unionised plants in industrial countries, to low-wage, unregulated and non-unionised production facilities in the developing world.

The economic power of multinational companies often also gives them tremendous political influence and means that they are able to undermine existing regulations, prevent new regulation or keep regulatory agencies at bay. Not that this is always necessary. Some countries, like Indonesia and Guatemala, have little meaningful occupational health and safety regulation. Whereas Mexico – which does have legislation – rarely enforces it.

The ILO estimates that, in the developing world, around one million workers die each year from illnesses caused by exposure to industrial chemicals. This has resulted in a surge in reproductive and neurological issues, respiratory and cardiovascular diseases and a range of cancers.

In the same regions, there is also regularly 'community exposure' associated with the operations of industrial facilities. Where there are inadequate resources to protect worker health, the surrounding area often suffers from contamination, either to the ground, air and water, meaning that it's not just the workers that suffer exposure to harmful substances, but great swathes of nearby residents are put at risk too.

The lax approach to safety legislation seen in many developing nations is often perpetuated not out of ignorance or greed, but out of necessity. Many of these countries are heavily in debt to institutions like the World Bank and the International Monetary Fund. As a result, they are dependent on foreign investment – and are, consequently, permanently vying for the attention of multinational companies. Consequently, anything that would stop investment looking like an attractive prospect – such as strict enforcement of occupational and environmental health laws – would be economically disastrous.

Even in the world's largest economy, however, there is limited improvement. Successive twenty-first century governments in the United States have failed to significantly update safety standards or create meaningful legislation to address newly-recognised hazards. The dramatic scaling-down of the workplace enforcement activities in the US clearly indicates that OSH is not currently a national priority.

The race to the bottom . . .

As the global 'race to the bottom'[2] continues, with the recorded number of unsafe and unhealthy workplaces growing, many workers across the globe are now being forced to compete in hiring themselves as minimum-possible-cost labour. As a result, in many of the poorest countries, workers are being forced to fight for jobs that are not just detrimental to their health but even, potentially deadly.

The recorded levels of toxic exposures have increased exponentially in the developing world, after a number of multinational companies repositioned their toxic work processes and waste sites in poorer nations from their previous locations within the developed world.

Back in the 1990s, globalisation was seen by most economists as an inevitable, unstoppable force. Since it takes many forms – including services, capital and ideas – globalisation has always been difficult to define. However, what it most commonly came down to is the process whereby organisations move their production sites from their affluent countries of origin (where labour is expensive), to poor countries (where labour is cheap.)

Multinationals relocating their operations away from the developed world, have seen a noticeable effect on their domestic markets too. Workers in the United States, for example, are now regularly reporting demands to accept pay-freezes – and even pay-cuts – from employers threatening to export jobs. This has naturally resulted in creating cultures of fear – as described by Frank Furedi and explored in the Introduction to this book – within many organisations.

. . . And the climb back up

In Western Europe, unemployment has grown hugely due to globalisation. Wages have also plummeted in the region – and, at the height of the 'global economic downturn' in 2008/9, nearly a quarter of the population, about 92 million people, were at risk of poverty or social exclusion.

In 2010, it was estimated that around 942 million working poor – nearly one-third of world's workforce – were living on less than $2 (USD) a day. It would only really be a matter of time before big business's growing reliance on cheap labour had consequences in high-wage areas.

With poverty, material deprivation and super-exploitation emerging even in the affluent parts of Europe, it means that the world is no longer a place where rich, well-educated countries and poor, badly-educated countries are neatly divided. Indeed, there are already stark divides in the health standards of populations of rich countries.

Let's take the United Kingdom as a case in point. A 2008 study conducted by the World Health Organization showed that the difference in average life expectancy between occupants of rich and poor neighbourhoods in Glasgow, Scotland – a city of some 1.2 million people – is a staggering 28 years – with most poor Glaswegian males not expected to see their 55th birthday (Reid, 2008).

Thanks largely to the interventions – and ensuing publicity – from agencies like the WHO, some international focus has been afforded to the 'social gradient of health' – with the United Nations agency demanding the gap is closed within a generation in the name of social justice.

A poignant example: Rana Plaza

On 24 April 2013, an eight-story commercial building named Rana Plaza collapsed in the Dhaka District of Bangladesh. The search for the dead ended a month later, with the official reports showing 1,134 confirmed fatalities. A further 2,500 workers were injured in the disaster. As the complex collapsed, the social shock wave rippled out across the globe.

The building contained clothing factories, a bank, a number of residential properties and several shops. However, whilst the shops and the bank on the lower floors had been closed and employees sent safely home the day prior to the accident – when workers reported seeing large cracks appearing in the foundations of the building – the clothing factory owners ignored their warnings and garment workers on the upper floors were ordered to return to work.

Following the disaster, the International Labor Rights Forum (ILRF), demanded that the multinational companies who had ordered clothing from the factories of Rana Plaza pay into a compensation fund for the victims' families, many of whom were struggling to survive having lost their main source of income.

However, of the 29 high-street fashion brands that sourced products from the Rana Plaza factories, only nine attended meetings to agree a proposal on compensation to the victims.

Several companies, including Walmart, Carrefour, Mango, Auchan and Kik, despite having made encouraging noises in the immediate aftermath of the accident, refused to provide any compensation. Others, such as Primark, Loblaw, Bonmarche and El Corte Ingles stated they would pay compensation – but were criticised for the small amounts and level of secrecy involved. It later transpired that fashion chain Primark were offering $200 (USD) compensation to the families – but only in the event that they could provide 'DNA evidence' proving that they were related to one of the bodies retrieved from the building.

Though the Rana Plaza disaster culminated in the largest manufacturing loss of life in history, it is far from unusual. The ILRF reported that from the beginning of the 1990s to 2012, at least 1,000 workers have died in fires and other building disasters in Bangladesh alone. This includes the Tazreen factory fire, also in Dhaka, which killed 112 workers at a textile plant in November of the previous year.

Following on from these tragedies, it emerged that the US and European companies, for whom the clothing was being created, had all publicised 'codes of conduct' that supposedly guaranteed well-maintained and safe working conditions throughout their entire supply chains. In practice, though, it meant very little. Rather than being legally required or enforced, the corporate codes are entirely voluntary. Local leaders chose to ignore them rather than invest financially in building maintenance, and the clothing giants turned blind eyes.

In any case, it transpired that the language of the 'codes' was also extremely vague. For example, many claimed a 'commitment' that workers along their supply chains will be given 'legal' – or, in some cases, just 'adequate' – wages. But, when these workers are residing in locations so far removed from the company's head offices, where there is no due process in place to investigate this (or legal incentive), what does this 'commitment' really amount to?

The retailer's codes of conduct also usually claim that they uphold the workers' 'rights of association' – or, rather, their entitlement to form a union. However, when they specifically choose locations in their supply chains where these 'rights' are either little understood, like Bangladesh, or entirely non-existent, like China, this seems somewhat disingenuous, at best.

Despite Bangladesh's appalling record on occupational safety and health, the factories in the country are booming. With 3.6 million garment workers, Bangladesh is the world's second-largest clothing exporter after China. Tazreen, Rana Plaza. Where next?

Ungoverned supply chains

Over the course of the last two decades, many large multinationals have been extremely forthright in their support of Corporate Social Responsibility (CSR). However, considering the huge death toll at the Rana Plaza – and the demonstrably poor response by those multinationals implicated – in many cases promoting CSR seems like little more than window-dressing – or, else, a way to pre-empt the argument, before genuine legislative measures are introduced.

In Primark's online presentation *Providing Customers with Ethically Sourced Garments*,[3] it states:

> Business has a responsibility to society. Business ethics are rules of conduct and principles and patterns of behaviour in business dealings that involve 'doing the right thing'. Part of this responsibility is to look after the wider community involved in the business process, including employees, across the globe.

These days, most multinationals will have sections on their corporate websites dedicated to making similar claims about their responsible business practices. However, if safety conditions are to improve, and further disasters to be avoided, there needs to genuine, meaningful regulation rather than Public Relations statements, voluntary codes of conduct and self-governing policies.

Whilst there are organisations that do invest money and resources into promoting ethical business processes throughout their supply chains, at the end of the day, it is never the priority – and in times of economic uncertainty, many companies, multinationals included, are driven to be as competitive as possible. Without proper regulatory controls governing all organisations, there is the very real danger that this will culminate in a race to the bottom.

Just prior to the events at Rana Plaza, one large multinational, The Walt Disney Company, the world's largest licenser, set out new rules for overseas production – and decided to withdraw operations from Bangladesh and Pakistan.

The multinational organisation had previously been dogged by international media reports about a string of factory fires within their factories in those countries. Evidently, the ensuing bad publicity made them rethink their supply chain, and being a publicly-held organisation, accountable to its shareholders – and one synonymous with children – Disney decided that, on this occasion, its reputation was more important than profits.

However, even regardless of possible reputational damage, ignoring, overlooking or not ensuring the health and wellbeing of workers in all parts of a supply chain seems somewhat myopic. Big brands exercise a huge amount of control on production within the factories they operate, so they have the power to positively influence working conditions. If they do not give compensation and partake in good-faith industrial relations, then the call for binding international legislation will only continue to strengthen.

Raising the game

In a global economy, if we are to have any chance of avoiding the seemingly inevitable slide downwards, and actually stop workers' lives being transformed into those of enforced drudgery in hazardous conditions, it is essential that research takes place that evaluates the health and safety costs of globalisation. Only when this has taken place, can international standards be created to effectively safeguard the health and lives of those individuals employed by multinational organisations.

Production processes would obviously need to be strictly regulated too, to ensure that whilst multinational organisations remain competitive, they are also sustainable, both economically and in terms of the lives of the individuals working within them.

Of course, it isn't true that all multinationals are operating sweatshops in low-wage regions, with hazardous working conditions and substandard wages – in fact, many do pay relatively high wages based on their location and offer higher labour standards. However, whilst there are examples being uncovered, the situation clearly needs to be addressed.

Globalisation has seen huge advances in productivity, opportunity, technological progress and in uniting the world's workers. However, worryingly, it has also seen an upsurge in poverty, deprivation, social disparities and violations of human rights.

THE NAKED TRUTH: GLOBALISATION

Over the last few decades, globalisation has seen many instances of multinational organisations moving their manufacturing plants to developing countries to capitalise on cheaper labour costs and avoid the more stringent health and safety legislation of developed countries.

Globalisation has had a profound impact on life in rich and poor countries alike, with most researchers citing it as the reason for such trends in the world economy as lower wages for workers, and higher profits, and low inflation and low interest rates, despite strong growth.

As globalisation continues, it is proving to be increasingly difficult to manage – and, as a result, remains largely ungoverned. Consequently, exploitation is on the increase: with many workers in developing regions being forced to fight for low-paid and dangerous jobs bereft of any kind of security.

Whilst the right to organise a union is enshrined in the resolutions passed by the ILO, the organisation lacks any real enforcement powers – and, in any case, in many locations, concepts such as unions, Occupational Safety and Health (OSH) and workers' rights are alien concepts – and workers, dependent on jobs to support themselves and their families, are simply powerless to effect any meaningful change.

For globalisation to sustain itself and be recognised as a force for good, it needs to help raise the rate of economic growth in developing countries and reduce world poverty. However, without the requisite legislation and robust enforcement powers, it is more likely to simply undermine the socio-economic security of poorer countries, erode human rights and result in a huge rise in work-related injuries and death.

GET NAKED: GLOBALISATION

1. How has globalisation affected your life? Do you think it has had a noticeable impact? If so, has it been largely positive or negative?
2. What impact has globalisation had on your job, organisation and industry sector? (Think of specific examples: this might be anything from how your personal remuneration or benefits might have been affected, to influences on the corporate structure, to recent changes in industry rules and regulations.)
3. Think of an organisation you know that has elements of its supply chain set up in a different country. What do you consider to be their motives for this? What factors might have influenced their choice of country?
4. Consider globalisation's influence on your country over the last few decades. On balance, would you say that the pros outstrip the cons? How has it affected the national outlook?

5 Where are the opportunities to create a more socially just and cohesive approach across your business globally? What do you need to do in order to avoid the 'race to the bottom'?
6 Primark states that "Business has a responsibility to society". Do you agree with this statement? How does your business ensure that it acts responsibly and ethically in all areas of operation – wherever these are across the globe? What messages do your organisation's codes of conduct, principles and commitments to 'doing the right thing' seek to communicate to your supply chain – and does your supply chain clearly understand what's expected of them in relation to these codes?

Notes

1 And, arguably, the corporate fascination for targets such as 'zero accidents' which often lead to mis- or non-reporting of workplace injuries, incidents and accidents. We'll explore this idea further later in this book.
2 'The race to the bottom' is a socio-economic expression describing a situation where organisations – or, indeed, countries – compete by cutting wages and living standards of workers – and where production is moved to locations where the wages are low, and workers have few rights. The phrase was coined by Adolf Berle and Gardiner Means (1932) in *The Modern Corporation and Private Property*.
3 www.scribd.com/document/191867180/Primark.

3 The legal landscape

In this chapter, we'll explore the horizon of international Occupational Safety and Health (OSH) legislation, and look at some of the regulatory frameworks adopted around the world. We'll also consider the history of safety law, and look at some of the events and legal landmarks that helped to shape it.

We'll explore how the regulatory bodies that oversee the legislation are instrumental in providing encouragement, instruction and enforcement, consider its principles and limitations, reflect on some of the factors that hold back safety legislation, and speculate on its future.

In the UK, protective legislation for workplace safety can be traced back as far as 1802, when the Act for the Health and Morals of Apprentices was introduced to stop orphaned children effectively becoming slaves to factory owners. However, it was in the 1830s and 1840s that the true foundations of regulatory control can be observed, with the enactment of statutes such as the Factories Act (1833) and the Fatal Accidents Act (1846).

These Victorian laws set the precedent for what was to come, and made way for arguably the most ambitious and robust regulatory framework for workplace safety ever devised – the Health and Safety at Work Act (HSWA) 1974. Since then the HSWA has been used as a model framework for many legislative texts around the world.

Based on key recommendations in the 1972 Robens report, the HSWA ensured the development of over 30 statutes and 500 sets of regulations and was described as "a bold and far-reaching piece of legislation" by workplace regulator the Health & Safety Executive's first Director General, John Locke.

It was a marked departure from the regulations previously in place, as it used a new goal-based, non-prescriptive model, supported by guidance and codes of practice. The new system was built on the premise that those who create risk are best placed to manage it – replacing a jumble of muddled industry regulations that had built up over the previous 100 years.

The HSWA enabled the creation of administrative bodies, outlined their powers of inspection and enforcement – and meant that, under the Act, British employers had a legal obligation to ensure that the health and safety of their workers is protected.

Whilst employers must, by law, have policies and procedures in place to ensure their workers' safety, this often has little bearing on how these policies are enacted or how strictly they are adhered to – which is more likely to depend on how robust – or weak – the 'safety culture' is within an organisation.

Whilst having a robust safety culture is not something that is legally enforced, following any notable or high-profile corporate accident, it is routinely the state of an organisation's safety culture that is scrutinised – and what regularly contributes to any prosecution, should there be one.

This has been illustrated a number of times in recent years following a number of high-profile cases. The King's Cross disaster, Piper Alpha and, more recently, Deepwater Horizon, Rana Plaza and Grenfell Tower have all highlighted the significant impact organisational factors have on safety culture. We'll explore safety culture in several other chapters of this book.

An important organisational driver for measuring health and safety performance is to decide whether risks to workers' health and safety are as low as 'reasonably practicable' – a concept that is fundamental to most of Great Britain's health and safety legislation. It is what underpins many of the requirements of the HSWA, the regulations enforced by the Health & Safety Executive (HSE), and by local authorities. Indeed, Section 2 of the HSWA imposes a general duty on employers to ensure: "so far as it is reasonably practicable, the health, safety and welfare at work of all his employees".

In particular, this duty extends to the provision and maintenance of safe plant and safe systems at work; safe use, handling, storage and transportation of articles and substances; provision of information, instruction, training and supervision; maintenance of safe place of work with safe access and egress; and the provision and maintenance of a safe working environment with adequate welfare facilities.

The law also recognises that unsafe work practices may impact not only those in direct employment – but, additionally, on those who may venture within its proximity. The HSWA imposes a duty on employers to safeguard persons not in their employ but who may be adversely affected by their activities – for example, visitors, customers, contractors and neighbours living or working in the immediate vicinities.

In a case that appeared before the Basildon Crown Court in 2004, a building firm was fined after they were contracted to replace some windows in a local school. In doing so, asbestos fibres were dislodged and contaminated classrooms. The principal contractor was fined under Section 3 (1) HSWA for failing to ensure the health and safety of non-workers – in this case the school pupils and teaching staff – and the building company was prosecuted for failing to provide a proper safety system for their own workers – a breach of Section 2 (2) (a).

The self-employed are similarly bound by the HSWA. In Section 3 (2), it states that the self-employed hold similar general duties to ensure that they – and other persons not being their employees – are not exposed to risks to their health and safety.

The legislation cuts both ways, so that the onus is not entirely on employers creating safe conditions for their staff. Under Section 7 of the HSWA, employees are directed to take all reasonable precautions and care of their own health and safety – and that of any others that might be affected by their 'acts or omissions' whilst at work. This not only means other employees, but has been interpreted in many court hearings to include members of the public also. Employees must also co-operate with their employers to enable them to 'meet their obligations', such as using equipment and systems provided by the employer in order to work safety.

Reasonably practicable

The term 'reasonably practicable' pervades the general duties within the Health & Safety at Work Act – and with just cause. Reasonable practicability is a concept that prevents the employer's duty becoming absolute. It is an expression concerned with the foreseeability of hazards and the costs of eliminating risk. Elucidating how they interpreted the term 'reasonably practicable', the Court of Appeal went on record with the following:

> Reasonably practicable is a narrower term than 'physically possible' and seems to imply that a computation must be made by the owner, in which the quantum of risk is placed in one scale and the sacrifice involved in the measures necessary for averting the risk, whether in money, time or trouble,

is placed in the other, and that if it be shown that there is a gross disproportion between them – the risk being insignificant in relation to the sacrifice – the defendants discharge the onus on them.

Where an employer provides measures to control risk, it is not sufficient to simply make these items available. The court case *Ginty vs Belmont Building Ltd* was actioned when a workman fell through an asbestos roof whilst at work. It was later revealed in court that the worker had not been using the crawling boards provided by the employer. The Court confirmed that the workman had to take some responsibility for his own safety. Nevertheless, it concluded that there was still liability in terms of the building company, due to its omission to provide direct instruction or training which it concluded went beyond 'the wrongful act of the employee'.

The duties imposed on employers by the HSWA create legitimate obligation. Although the duties are not absolute, they fall short only by the expression 'so far as is reasonably practicable'. In the event of an accident, the prosecution needs only to prove that the defendant was an employer, thus triggering the statutory duties, and that beyond reasonable doubt some foreseeable risk existed. The onus of proving that all reasonably practicable steps were taken falls to the employer. If they are unable to do so, they will be convicted and are then vulnerable to an unlimited fine in the Crown Court.

As an 'enabling' Act, the HSWA provides a framework for subordinate legislation to be implemented, adding detail to the statuary principles by clearly defining duties pertinent to specific risks or industries. Paradoxically, given Lord Robens' preference to avoid too much law, there has followed a significant tranche of regulation, much of it driven by the European Community in the last few decades.

Management of Health and Safety at Work Regulations

The employer's duties set out under HSWA are further developed by the Management of Health and Safety at Work Regulations (MHSW).

This secondary legislation was drafted in Great Britain to the meet the requirements for the European 'Framework Directive', which, in 1989, introduced measures to encourage improvements in the safety and health of workers at work. Other European Member States drafted and invoked similar local laws to meet Directive.

The Management Regulations revolve around the concept of risk management, introduced to the employer through Regulation 3, which states that: "Every employer shall make a suitable and sufficient assessment of the risks to health and safety of his employees to which they are exposed whilst they are at work."

This regulation requires employers to conduct a systematic assessment to identify workplace hazards and evaluate the risks involved, in order to introduce appropriate control measures. The significant findings of the assessment must be recorded, either electronically or in writing, if five or more persons are employed

– and reviewed and modified whenever there is a "significant change" or where there is "reason to suspect that (the assessment) is no longer valid".

Whilst the requirement to conduct risk assessments has existed for many decades, the Court of Appeal recently felt it pertinent to re-emphasise the importance of their adequacy:

> The whole point of proper risk assessment is that an investigation is carried out in order to identify where the particular operation gives rise to any risk to safety and, if so, what is the extent of that risk . . . and what should be done to minimise or eradicate that risk.

Further examination of the law reveals it is not seen as appropriate for employers to rely on the experience of the workers or for them to exercise 'common sense' or act intuitively. Instead, a 'competent person' must be appointed to exercise a methodical approach to workplace safety management.[1]

When considering the obligations for competency and assessment together, it is evident that the implications of the Management Regulations are particularly far-reaching for certain industry sectors – for example, the construction industry, where the workplace may be a vast building site which changes rapidly as work progresses. The challenge is further compounded when one considers the scope of the assessment. Workplace hazards, such as fire, noise or asbestos must obviously be included, but more subtle risks, those arising from the work process itself, such as manual handling or the creation of fumes or dust, must also be considered. In recent years, the courts have also added intangible factors, such as psychological stress from overly-demanding work schedules, harassment and workplace bullying into the mix.

Potency of the law

As the foundation stone of British health and safety law, the HSWA imparts general duties which are then expanded by regulations – such as the Management Regulations described above, and other defined laws related to particular risks – to add specific detail. When taking enforcement action, it is commonplace for the prosecution to prepare an argument based on a general duty breach together with specific secondary legislative failings.

In certain industry sectors, such as within the construction industry, there may be a number of duty-holders involved, such as contractors, sub-contractors, designers and supervisors. Accordingly, a flexible approach to prosecution is crucial in order to maximise the potency of the law.

The case of *R v Lindsay Barr and Others* illustrates this legal flexibility well. Principle Contractor, O'Keefe and Co sub-contracted Britain Construction Ltd to conduct excavation works on site for which Lindsay Barr, a self-employed consultant engineer provided structural advice and drawings. When a section of excavation collapsed, fatally injuring an employee of Britain Construction, the court prosecuted Barr under the HSWA for failing to conduct his undertaking in a way that did not expose persons to risk; O'Keefe & Co. were also prosecuted under the HSWA for failing to ensure the co-ordination and cooperation of contractors, and, again, for failing to ensure the stability of structures. Britain Construction were then found to be in breach of the HSWA Section 2(1) for failing to ensure the safety of their employees, and Regulation 3(1) of the MHSW for failing to conduct an adequate risk assessment.

The phrase 'reasonably practicable' theoretically turns the obligations under HSWA and regulations into duties of reasonable care. In practice, since the burden of proof of practicability rests on the defendant, the true effect may be to impose absolute liability, with proof of fault.

Despite nearly 40 years and a radically transformed corporate landscape, the HSWA has stood the test of time. During its most recent review, *Reclaiming Health and Safety for all* (Löfstedt, 2011) – written at a time when the then-UK Prime Minister, David Cameron, was complaining about the 'health and safety monster' and asserting that he wished to 'kill off safety culture' once and for all – Professor Ragnar Löfstedt concludes with the following line: "In general, there is no case for radically altering current health and safety legislation."

Some issues have persisted, however – even in spite of the HSWA. In the fields of construction and agriculture, for example, little has changed since the Act's introduction. On its enactment, former Director General of the HSE John Locke said the following:

> In construction, the same basic causes have produced a high proportion of accidents in the past 60 to 70 years: most of these accidents happen to people engaged in routine site activities which simply have not received sufficient forethought and care.

Sadly, this has changed very little over the ensuing 40 years – and Locke's comment holds true today.

The Löfstedt report led to some streamlining of health and safety regulation in the UK. Ultimately, David Cameron got his wish to relieve the 'burden' of health and safety on British businesses, though possibly not to the extent he might have wished.

Brexit

Many of the day-to-day responsibilities of individuals working in health and safety roles stem from legal requirements set by Europe.

It is not certain that such regulations would remain in force following 'Brexit' – the UK's departure from the European Union. However, it is probably worth remembering that during the campaigning, which resulted in a positive outcome for the 'Leave' team, one of their key stipulations was to reduce the regulatory burdens on commercial organisations caused by Europe.

When it comes to occupational health and safety, though, drastically downgrading current legislation for workers residing in the UK would certainly be problematic, as they have come to expect certain rights and standards.

In any case, the Europe Union is now independently going through the process of simplifying and reducing compliance requirements – and has already invested considerable resource on deregulation. The continuing Regulatory Fitness and Performance Programme (REFIT) is charged with reviewing all EU regulations to see which can be removed or consolidated. Consequently, in the decade since 2006, more than 6,100 acts have been either withdrawn or replaced.

In 2014, the European Commission published a new Strategic Framework on workplace Health and Safety setting out its priorities for the coming six years. Though it is currently unclear what the UK will do post-Brexit, assuming it will wish to remain competitive with the rest of Europe, it probably makes sense to explore the focus of the EU's future activities.

The Strategic Framework states that the three major challenges for the EU will be to improve implementation of existing health and safety rules – especially among micro and small enterprises. They also wish to address work-related diseases and explore how to improve care for the ageing EU workforce.

The framework also outlines the following key strategic objectives:

- Further consolidating national health and safety strategies through, for example, policy coordination and mutual learning;
- Improving enforcement by Member States by evaluating the performance of national labour inspectorates;
- Simplifying existing legislation where appropriate to eliminate unnecessary administrative burdens, while preserving a high level of protection for workers' health and safety;
- Improving statistical data collection to have better evidence and developing monitoring tools;
- Reinforcing coordination with international organisations – such as the International Labour Organization (ILO), the World Health Organization (WHO) and the Organisation for Economic Co-operation and Development (OECD) and partners to contribute to reducing work accidents and occupational diseases and to improving working conditions worldwide.

Beyond the UK

In the UK, and in other developed countries, workplace health and safety is typically planned and budgeted for as an everyday part of organisation's operational life – and workplaces have regulations and intricate systems of enforcement, designed to suitably encourage businesses into providing safe work environments for their people.

This is not the case the world over, however, and as we've seen in the chapter on globalisation, there is the opportunity for international commercial corporations to seek to exploit the lax safety regulations in developing countries in order to maximise profits.

At this point in our discussion, it may be interesting to take a view of the different safety models and types of legislation that exist around the globe –

if for no better reason than to understand how far we've come – and what might still might be achieved.

United States of America

In the United States, the Occupational Safety and Health Act passed into law on the 29th of December 1970. The Act created the three agencies that administer it – the Occupational Safety and Health Administration, National Institute for Occupational Safety and Health and the Occupational Safety and Health Review Commission.

The Act also sanctioned the Occupational Safety and Health Administration (OSHA) to regulate private employers in the 50 states, and made a legal requirement for all employers to provide employees with "employment and a place of employment which are free from recognised hazards that are causing or are likely to cause death or serious physical harm to his employees". Note the similarities with the HSWA.

The OSHA's official headquarters are in Washington DC, but there are now an additional ten regional offices, each of which is organised into three sections promoting compliance, training and assistance.

It was initially intended that the OSHA would administer 50 state plans with OSHA funding. However, almost half a century later there are only 26 approved state plans – of which, only four cover public workers. The rest of the states have declined to participate in the programme.

The OSHA creates safety standards – the Code of Federal Regulation – and helps to enforce this through inspections conducted by Compliance Officers. Worksites can apply to enter OSHA's Voluntary Protection Program (VPP), the application of which necessitates an on-site inspection.

If the worksite passes the inspection, it is then granted VPP status and OSHA doesn't visit the site again for three to five years (when the VPP is reviewed again), unless there is a work-related fatality, or a worker makes a formal complaint about the organisation. Though this seems quite hands-off by European standards, VPP sites have injury and illness rates that are less than half industry averages.

The OSHA also produces a range of health and safety publications, provides advice to workers and funds training for small businesses. Its Alliance Program helps safety teams of organisations to develop compliance tools and resources, share information with workers and instructs them on specific industry rights and responsibilities. Most of the focus of the OSHA is on small and micro businesses and high-hazard industries.

Middle East

Most major organisations in the Middle East are allied to businesses in developed nations where workplace health and safety is taken seriously. In places like Saudi Arabia, Bahrain and Qatar, international corporations such as Exxon Mobil, Shell and BP essentially partner with local ministries to create work projects and corporations. Aramco, the Arabian American Oil Company, for example, has world-class safety systems, since they simply brought their safety management systems with them from the US, and then employed local workers to run them. By contrast, workers at home-grown organisations in these nations are not afforded anything like the same level of workplace safety.

Asia

In India, the focus of occupational safety and health is mainly based on factory and dock workers. The Labour Ministry devises policies on occupational safety and health with assistance from the Directorate General of Factory Advice Service and Labour Institutes (DGFASLI), and enforces these with regular spot checks by inspectors. As the technical arm of the Ministry of Labour & Employment, the DGFASLI also delivers technical support, conducts safety surveys and creates occupational safety training programmes. More recently, it has also begun to instruct factory workers on matters of workplace health, efficiency and wellbeing.

China has two regulatory bodies responsible for occupational safety and health: The Ministry of Health, responsible for workplace disease prevention, and the State Administration of Work Safety which deals with matters of workplace safety. These have been set up to enforce new government acts that were ratified as recently as 2002 – the Occupational Disease Control Act of the People's Republic of China and the Work Safety Act of the Peoples' Republic of China. The Occupational Disease Control Act is still being considered – and, though it represents a move in the right direction, is evidently not a priority.

Singapore's Ministry of Manpower vision is: 'A Great Workforce; A Great Workplace'. The government department, which used to be known as the Ministry of Labour until 1998, is responsible for all policies relating to Singaporean workers. As well as overseeing matters related to immigration, visas, grants of citizenship, etc., the ministry is also in charge of safety, and, as such, operates checks and campaigns against unsafe work practices. Like the US, much of the focus of this safety work is centred on workers in high-hazard jobs – such as those working at sea, working at heights and those involved in traffic management.

The impact of politics

The local political climate can have a huge impact on workplace safety. Probably the most extreme example of this is in Iran, where the governing council is answerable to a panel of theologians. Therefore, all laws and regulations must be sanctioned by the Iran Guardian Council to ensure they conform to Islamic law. Since most proposed safety regulations are perceived as being intrinsically outside Islamic tradition they are therefore not supported by the government.

Romania and Kazakhstan, who are still living in the shadow of huge social and political upheaval – and where workplace safety has traditionally never been seen as a priority – are now seeing a flood of international companies, such as Shell, BP and Chevron, setting up in the country. Consequently, the imported work methods of these organisations are now beginning to show significant influence throughout industries there – and have a positive effect on safety locally.

Another major influence on workplace safety within the developing world is, of course, poverty. When workers are facing starvation, naturally their priorities change. Choices that are open to workers in the developed world, are rarely feasible to those in the developing countries. After all, the motivation in sustaining their lives naturally overshadows the vague threat of potential future injury. Workers may feel compelled to accept work that is inherently risky and to work for companies that do not provide safety training or equipment. Governments of Third World countries and companies that invest in Third World projects must be consistent in their insistence on improved workplace safety conditions to have a long-term impact on improving workers' safety. Unfortunately, corruption still plays a large role in many of these countries, or at least within the governmental and/or corporate structures of these countries.

Some countries are hit by a range of adverse factors. The Central African Republic, for example, is not only hit by poverty and social upheaval, but was recently torn apart by civil war. The country's Ministry of Labor and Civil Service has never actually defined health and safety standards in the workplace – never mind enforced them. As a result, workers have no security, working conditions are harsh and child labour is rife.

Workers demanding better conditions in Cambodia are subject to unfair dismissals, intimidation, arrest and even violence. In January 2017, Cambodian police opened fire on garment workers demanding a higher minimum wage at a protest in Phnom Penh, killing three people and wounding more. Several protesters were arrested, and others were beaten by police at separate rallies demanding better wages during the same year. At least 500 workers fainted at garment factories in 2017, and more than 4,000 have fallen unconscious at work in the past four years.

The legal landscape

The UK has long been recognised as being at the forefront of Health and Safety at Work regulation since the practice began. The HSWA has become regarded as one of the leading pieces of health and safety law globally, and is a beacon for excellence that has been emulated by many other nations around the world. As Brexit negotiations continue, the legal landscape is likely to remain to be something of a voyage into the unknown for some time to come. However, it is unlikely that UK politicians will have the appetite to laboriously dismantle existing health and safety law. The UK is likely to continue with its successful risk-based health and safety system (including laws from EU Directives), which has been found to be robust and fit-for-purpose by a number of independent investigators. As time marches on, Brexit is much more likely to affect future law-making and implementation.

We can anticipate some tidying up and consolidation of EU health and safety regulations to take place post-Brexit – after all, it is something that the EU are doing themselves currently. However, it is unlikely that UK health and safety law would be subject to drastic change, especially if the UK remains in the EFTA (European Free Trade Association) and the EEA (European Economic Area).

There are some concerns raised by safety practitioners that health and safety and equality laws may see a negative impact post-Brexit. Whilst the HSWA was created in the UK, and there was some additional legislation regarding discrimination in place before the UK became a part of the EU, the lion's share of these laws stem from EU directives and European courts. Many rights that workers enjoy now were simply not in place before the UK joined the EU. For example, employers had no obligation to undertake health and safety risk assessments, there was no universal right to paid holidays, part-time workers enjoyed very few workplace safety provisions and there were no laws barring discrimination on the grounds of sexual orientation.

So, what will the real impact of Brexit on workplace health and safety be – on the United Kingdom, and more broadly around the world? This chapter has plotted the development of workplace health and safety law and enforcement, and tentatively explored the legislature further afield. We've noticed that the Health and Safety at Work etc. Act of 1974 provides a good example of solid workable law that has been emulated by many other nations. It is likely that any significant change to UK laws, post-Brexit, will have at least some impact on both developed and developing nations around the globe.

THE NAKED TRUTH: THE LEGAL LANDSCAPE

Occupational Safety and Health (OSH) legislation is crucial in ensuring workplace safety standards – and the continued health, safety and wellbeing of workers.

At its most comprehensive and effective – as with the UK's Health and Safety at Work Act 1974 – a legal framework for organisational safety is vital in defining the obligations of employers, workers, contractors and the individuals that control, operate or maintain workplaces. It also has extensive powers of assessment, review and enforcement, being ultimately backed by the ability to bestow punitive measures for non-compliance – such as sanctions, fines and, in extreme cases, imprisonment.

Looking beyond the legal imperatives, adhering to a tight safety framework needs to be thought of a positive step. Not only does it promote fairness, by introducing industry– and nation-wide standards of care, it actually promotes organisational performance, improves sustainability and creates success. By investing in the health and safety of workers, employers boost morale, reduce accidents and work-related ill health and drive efficiency. In doing so they stand a chance at not only cutting accident losses but also increasing profits. We'll explore how this occurs in subsequent chapters.

GET NAKED: THE LEGAL LANDSCAPE

1 How does health and safety legislation affect your organisation? Has it led to an increase in sensitivity regarding safety? Or higher levels of anxiety when it comes to risk? Why would this be so?
2 What is the impact of workplace health and safety legislation on everyday organisational practices and behaviours?
3 How do you feel personally about how health and safety rules are enforced in your business?
4 How is information about health and safety legislation communicated by the leadership team? Are workers at all level able – and encouraged – to report violations? Are the channels of communication adequate? Are the responses?
5 In your view, how do you rate your organisation's leadership when it comes to interpreting and administering health and safety legislation? Has a fear of litigation become a key driver of organisational behaviour? Why do you think this is?

Note

1 We'll explore the notion of a 'competent person' in a later chapter when we consider the role of the health and safety practitioner.

4 Governance

Governance is the systematised approach through which organisations are directed, controlled and assured – typically by their Board of Directors, though certainly elements of governance filter down through the organisational hierarchy.

In stripping things back in this book, governance is distinguished from both leadership and management. That is not to say that neither leadership nor management have responsibility for governance. The former may contain aspects of mid-range governance – particularly with regard to the direction of travel for the organisation and its strategic aims and objectives, and the latter may involve ensuring short-term governance through day-to-day control of process and activity and the everyday decisions required to keep an organisation operating efficiently and effectively.

In this chapter, we'll explore what governance really is and why it's essential to modern organisations. We'll also look at the relationship between governance and safety leadership, risk management and personal accountability.

The conviction of Don Blankenship, Massey Energy's Chairman and CEO – following the deaths of 29 miners employed by the company – is a stark reminder of the key role that senior executives play in safety governance.

A federal grand jury indicted Blankenship on the 13th November 2014, for conspiracy to violate mandatory federal mine safety and health standards, conspiracy to impede federal mine safety officials, and for making false statements to the US's Securities and Exchange Commission.

The case emphasised what can happen when an organisation's senior leaders lose sight of effective safety leadership because they're distracted by a focus on financial matters.

The responsibility of governance

Good governance is fundamental to the health of any modern organisation. However, there are still many organisations that don't really understand what it means in practice, or how to go about trying to ensure it actually happens.

There seems to be a number of basic misunderstandings when it comes to corporate governance and as a result, many business leaders look on it as a largely theoretical concept. This is not the case – governance is eminently practical.

It not simply a list of lofty ideals to be bandied about the boardroom every once in a while; governance should be recognised for what it is – a business activity, which, when conducted properly, should pervade every corner of the organisation, from senior leadership right down to the shop-floor.

What is governance?

Governance in the broadest sense is the process by which organisations are directed, co-ordinated and assured by their Board of Directors (BoD).

It works at a higher level than leadership vision and strategic activities, and beyond the realm of management, which typically concerns itself with the everyday operational decisions and actions required to run the organisation.

Governance relates to those processes by which senior leaders and managers are held to account and through which the broadest strategic decisions are taken. It is not simply about understanding what is the correct way to act in certain situations, organisations need to also demonstrate that they are delivering on these understandings.

There are frequently assumptions that once a top-level operational decision has been made it will automatically permeate throughout the rest of the organisation – but the reality is that it takes considerable time and resources to drive any practical change.

There is also often confusion within organisations as to where governance sits – and who is responsible for it. Often, front-line workers will consider it as little more than a set of 'Best Practice' guidelines created by individuals with little knowledge of the actual operational workings of the organisation.

In many cases, this might be forgiven – as it may not be all that far from the truth. After all, board members are not omnipresent entities – whilst they should have a good understanding of what goes on within the organisation in general terms, they will not be able to see everything that happens there. It is for this reason, that good governance needs to be entirely inclusive – and why everyone within the organisation needs to be responsible and held accountable. If workers – at any level – observe some part of governance is being ignored, this needs to be reported.

Before this can happen, however, suitable lines of communication need to be in place. Not just so that workers are aware of what is stipulated in the governance, but, also, so there are suitable channels available to them to make the reporting of failures possible.

Governing safety

Safety governance is about ensuring that the most senior leaders within an organisation have the knowledge, tools and processes in place to boost the safety performance of the organisations they lead to an area beyond mere compliance and safety legislation and gain positive assurance that compliance is achieved, and generally that things are on track.

Governance concerns how the BoD and Senior Executive team of an organisation conduct 'safety leadership', and the structure through which the commitment to their safety vision is set. It is also through governance that safety objectives are established and the process for monitoring performance is decided.

In *A Study of Safety Leadership and Safety Governance for Board Members and Senior Executives*, Dr Kristin Ferguson states that understanding the 'maturity' of safety governance within organisations is essential to progressive safety leadership. Ferguson identifies the following five stages of safety governance maturity:

- Transactional
- Compliance
- Focused
- Proactive
- Integrated.

Transactional – The BoD and Senior Executive team are demonstrably more concerned with production rather than safety – and ignore safety issues, viewing them as something under the purview of the Safety Team. They will typically only become engaged on the topic of safety after an incident has occurred.

Compliance – Compliance with health and safety legislation is the main driver of reporting to the BoD and Senior Executive team. Focus is on ensuring the legislation standards are satisfactorily met.

Focused – The BoD and Senior Executive team regularly ask detailed questions about safety issues and are interested in the causes of incidents and regularly take part in site inspections in order to further their understanding.

Proactive – The BoD and Senior Executive exhibit a strong understanding about safety and safety culture and recognise it requires safety leadership both inside and outside the boardroom.

Integrated – The BoD and Senior Executive team actively seek to understand the safety impacts of every decision made across the organisation. The concept of 'safe production' sets the tone for all health and safety discussions.

According to Ferguson "every organisation moves along a continuum as they develop a safety governance framework". Plotting where they are currently placed on the scale, often assists organisations to recognise their own failings – and, in doing so, helps them on their journey towards maturity.

When it comes to Occupational Safety and Health (OSH) governance, it's clear that there is no 'one size fits all' solution – the shape of governance will naturally depend on the types of OHS issues and the size and complexity of the organisation involved.

The Health & Safety Executive, Great Britain's workplace regulator, has outlined a neat framework for what Best Practice in OHS governance should look like by identifying seven basic principles (each of which chime well with Ferguson's study). Let's explore these now:

- **Director competence** – Directors need be fully aware of the main OSH issues relating to the organisations they run and remain up-to-date with any significant developments.
- **Director roles and responsibilities** – All Directors should appreciate the legal responsibilities of their roles in relation to governing OHS issues relating to the organisation. This must include terms of reference, OSH policy and strategy, setting standards, worker performance monitoring and internal control.

  At least one nominated Director should be personally tasked with overseeing and challenging the organisation's OSH governance process.
- **Culture, standards and values** – The BoD should take ownership for key OSH issues, uphold core values and be ambassadors for Best Practice.

  They should set the correct tone at the top of the organisation and establish an 'open culture', so that workers at all levels are able to report instances of bad practice.
- **Strategic implications** – The BoD should be responsible for driving the OHS agenda, understanding the risks and opportunities that might come about due to market pressures and proactively seek out anything that could compromise proper operations – and, ultimately, create and implement strategies to 'future proof' them.
- **Performance management** – It is the BoD's responsibility to set out the key objectives for OHS management and create an incentivised structure for senior management to drive good OHS performance.

This needs to balance both leading and lagging indicators and involve both tangible and intangible aspects.

- **Internal controls** – The BoD needs to ensure that OHS risks are managed adequately and that compliance with the core standards is met. Essentially, governance structures should enable management systems, actions and levels of performance to be challenged. As much as possible, this needs to use current internal control and audit structures.
- **Organisational structures** – The BoD will integrate the OHS governance process into the main corporate governance structures within the organisation – including committees overseeing risk, remuneration and audit.

Governance up, risk down

Recent research on governance by the business consultancy McKinsey & Company have shown that Directors are becoming increasingly confident in their knowledge of the organisations they serve – and becoming more tactical in their approaches.

According to the company's latest research, out of the 772 global organisations that responded to the survey, the figures show there has been a sharp rise in the amount of time boards spend working on strategy, since the previous McKinsey survey was circulated two years earlier.

Directors also seem to be more comfortable in their roles since the previous survey – and believe themselves to have more comprehensive understandings of the various OHS issues. In the most recent survey, approximately one-third of the Directors interviewed claimed to have a 'complete understanding' of current strategy, compared to roughly one-fifth two years previously.

However, whilst the number of Directors getting involved with strategy has obviously increased exponentially in a very short period, the amount of time devoted to Risk Management actually went down.

The area where most Directors indicated that there was room for improvement was with risk – with many stating that their boards regularly 'struggle to comprehend' – and (rather worryingly) 'make time to effectively manage business risks'.

It is not quite clear why organisations have become complacent to risk – but it does seem, on the face of it at least, to be something of a disturbing global trend.

An organisation's Board of Directors really doesn't have the luxury of being myopic when it comes to risk. In the event of a disaster befalling an organisation, the repercussions can be devastating.

Whether the costs to the organisation are reputational, financial or at the expense of the health or lives of the individuals working there, when it comes to investigating the causes of the accident, at least partial blame is likely to be attributed to the individuals responsible for putting in place strategies that might have avoided it.

In any case, what is governance for if not to protect the interests of the organisation? By cutting down the amount of time spent on risk management, it implies that Directors are simply burying their collective heads in the sand and just hoping for the best.

Accountability

The seventeenth-century playwright and humourist, John Baptiste Molière, once declared "It is not only what we do, but also what we do not do for which we are accountable." When it comes to safety, it's hard to imagine a more cogent remark – serving, as it does, not only as an accurate statement of fact – but, also, a stark warning.

Understandably, when it comes to safety governance, 'accountability' is not always a popular and immediately-welcomed concept; concerned as it is with encouraging and attributing liability.

It is also not necessarily that straightforward either. When something goes wrong within an organisation the focus of responsibility may not be easily apparent. Certainly, in larger organisations, when an incident takes place, there are frequently so many factors – and factions – involved it can be challenging to identify causal reasons.

In *Restoring Responsibility* (2005), political ethics pioneer, Dennis Thompson, refers to this as the 'problem of many hands'. Thompson believes that the way that most organisations are structured – with many individuals contributing to decisions and policies – creates real problems in terms of accountability.

It tends to result either in a worker being unfairly held accountable – and punished – for an incident which, in reality, they may have only had a very minor responsibility for; or in a worker symbolically 'accepting responsibility' for the incident in an effort to draw the safety investigation to a conclusion, without suffering any consequences. Or, otherwise, the entire organisation is held accountable – which is, of course, meaningless and platitudinal.

What might make more sense would be to broaden the criteria for individual responsibility, so that workers are held accountable for not anticipating failures in the organisation.

Ultimately, accountability needs to evolve to focus on active safety performance (the 'inputs' to creating safety), rather than latent safety results (the 'outputs' such as accident frequency metrics).

Restoring responsibility and aligning accountability

So, how to make the shift?

One way to get started is by embedding structured risk discussions into management processes throughout the organisation.

The Board of Directors and Senior Management team need to develop accountability, so that it drives excellence in safety performance. A methodological approach should be created to ensure that individuals are receiving consistent reinforcement of expectations, and are held accountable for the vital performance necessary to succeed. When there is no accountability, workers tend to underperform for no better reason than they don't know what they are meant to do.

In some organisations, it is the practice to link accountability to safety lagging indicators – and then punish workers for getting injured. This is not a sensible plan – as it contains very little incentive for workers to report an issue – and naturally drives reporting down. This means that safety issues remain, and problems endure.

It's essential that accountability is based on performance, not results. Instead of reprimanding workers for things they've failed to do, it would make more sense to introduce a more positive approach. Managers need to encourage workers to focus on a series of specific small improvements to safety performance, and then increase these incrementally over time – in order to drive sustainable safety performance improvement.

Safety leadership

When building and implementing new safety programmes, many organisations overlook one of the greatest safety management tools they have in preventing injury and accident – safety leadership.

Effective safety leadership can spell the difference between an organisation having an average safety record and industry-leading performance.

We'll explore safety leadership in more detail elsewhere in this book, but for now, with regard to governance, we can consider five key components of effective safety leadership that can really help generate a sense of positive governance. Let's explore them in turn now.

- **Visibility** – It is essential that safety leaders are visible and engage with workers. Taking time out to walk about the organisation and asking workers for their thoughts and ideas about potential safety issues will not only give leaders the 'heads up' on possible issues – but also demonstrate to workers that they – and, by extension, *the organisation* – value their contribution (and their safety).
- **Communication** – Effective safety leadership pivots on good communication skills. Without interaction with workers and strong communication, it's difficult for safety leaders to drive worker behaviour, because much of this depends of sharing the 'safety vision', building trust and providing positive role-modelling.
- **Feedback** – For safety leadership to be effective, it needs to be inclusive and touch every part of the organisation. Avenues of communication between workers and leaders need to be established, prominent and easily accessible. Rather than stigmatising the reporting of potential safety issues, this needs to be actively encouraged by all leaders and managers, organisation-wide.

- **Accountability** – If accountability is to be meaningful in an organisation, it starts with the leaders. There is nothing more likely to ferment distrust and unrest within an organisation than double standards. Leaders need to lead by example, not be exempt from the rules.
- **Innovation** – Good safety leaders understand the safety doesn't stand still, it's on a permanent mission of renewal and self-improvement. Strong leadership recognises the importance of trying out new processes and ideas in its never-ending quest towards zero accidents. Establishing a continuous improvement roadmap will also help ensure that safety programmes and systems keep pace with the rate of change in the rest of the organisation.

Drivers for governance

Top tier leaders are not all cut from the same cloth. There is not a single 'type' of Board Director and some may be far less concerned about governance than others.

It can be common in some organisations where Board Directors have become more focused on financial or production matters, and – usually following a period with few notable safety incidents – they grow confident that safety pretty much 'takes care of itself'. (They will rarely readily admit this, of course, but dig a little deeper and it becomes clear.)

However, there are significant benefits to strong safety governance that should be of interest to all Directors, regardless of organisation or industry sector. These benefits can be broadly categorised as moral, legal and financial.

Moral

As we discussed in an earlier chapter, all organisations need to operate in adherence with specific legal obligations – but more than that, they also operate within society, which imbues further responsibilities.

These societal obligations manifest as a moral responsibility to provide a 'duty of care' for its workers, customers, the local environment or anything else that might conceivably be negatively impacted by the operation of the business.

Organisational leaders have a moral obligation not to abuse the trust placed in them by broader society – and when they shrink from societal concerns (whether external in the community or internal with the workforce) over health and safety or when doubt is raised over the discharge of their basic duty of care, it can be detrimental to their public image and reputation. Such reputational damage is often followed by disassociation, a loss of business and drop in market value. Following the explosion on the Deepwater Horizon oil exploration platform on 20th April 2010, BP's market value plummeted more than 30 per cent overnight. And subsequent to the multiple fatality at their plant in La Porte, Texas, the DuPont corporation – once heralded the world's safest company, experienced a significant lack of market trust in their services after the Head of the United States Chemical Safety Board publicly suggested that the company could not manage safety properly within its own purview.

Legal

Though legal issues are typically the focus of an organisation's leaders, rather than that of the Board of Directors, following the prescribed set of industry rules and regulations should be such a significant driver for any organisation that they need to form the basis of any credible governance framework.

Not only do organisations that operate legally and safely tend to retain a strong public profile, any organisation that wants to be efficient and cost-effective (which is presumably the majority of organisations around the world) needs to do as much as they can to assure legislative compliance simply because prosecutions and the impositions of penalties waste valuable time and resources that might otherwise be put to better use.

Additionally, in many industries, a good compliance record is vital in maintaining the 'licence to operate'.

Financial

Though some may find discussing personal safety and finance in the same chapter discomforting, there are clear and strong financial drivers to encourage Boards of Directors to take responsibility for safety governance.

In many sectors – most notably the private sector – investors have been actively influencing board action by promoting the demonstrable links between good governance and lasting shareholder value.

Health and Safety Indicators for Institutional Investors (Mansley, 2002), a review of institutional investor policies by organisational strategists The Claros Group shows that safety policies and strategies are something that financiers are now taking a great interest in before investing. It's clear that in a turbulent and crowded market place, investors now see a solid approach to safety governance coupled with a good safety record as a strong differentiator.

Some of the things they look out for include:

- a strong public image and lack of adverse publicity;
- bottom line savings, based on low premiums, fines and worker absenteeism;
- a lack of worker and union pressure;
- an engaged workforce that contributes to the success of safety at work.

Research by Barclays Global Investors (2005) agrees that a strong safety record is now a major influencer on investment as it is a sign of a healthy, well-managed and durable organisation.

Considering the current uncertainty in the marketplace, investors are evidently looking for all the assurances they can find.

THE NAKED TRUTH: GOVERNANCE

Due to the prevalence of high-profile disasters of recent years, organisational safety is attracting more publicity than ever before. Consequently, the costs of getting safety wrong are becoming increasingly damaging. Increased publicity means that safety problems now have a greater negative impact on an organisation's reputation – which is often the catalyst for a significant loss of value.

Even a cursory search reveals that the internet is now littered with tales of corporate woe for organisations as diverse as sports clothing brands, chemical giants, construction firms and theme park owners following workplace health and safety failings. Each and every one tips the business upside down and causes revenue, market share, reputation, societal trust and employee morale to plummet.

Since an organisation's Board of Directors is responsible for developing and administering the governance framework to safeguard against such negativity, and as a Board are answerable to stakeholders, averting safety problems should certainly be one of their primary responsibilities. As such, safety must receive the same focus and attention as production figures or financial performance.

The responsibility of Directors does not stop with the issuing of strategy and policy directives though – it also needs to ensure that these are deployed effectively. A robust health and safety culture, supported by a visible and impactful governance framework, will allow managers and workers to take individual ownership of safety issues – which beyond providing assurance to the Board will also result in improvements to operational performance.

GET NAKED: GOVERNANCE

1 Look again at Ferguson's five stages of governance: *Transactional*, *Compliance*, *Focused*, *Proactive* and *Integrated.* Which stage do you think your organisation is at? What evidence do you see that suggests this?
2 What barriers stand in the way of progressing beyond this level of governance? And what are the positive enablers for further progression?
3 What's the strongest driver for governance in your organisation? Is it the moral imperative – keeping people safe from harm because it's simply the right thing to do. Or is it protecting the financial aspects of the organisation from harm. Or does a fear of adverse publicity and legal action drive your Board in its approach to governance? How do you feel about this? What would the workforce say?
4 Molière said that "It is not only what we do, but also what we do not do for which we are accountable." Where are the gaps in your organisation's approach to governance? What risks can you foresee that need to be woven into the way your Board is assured of organisational progress?

5 Consider the five key components of good safety leadership: *Visibility*, *Communication*, *Accountability*, *Feedback* and *Innovation*. How would you rate your organisation's Board in each of these? What could you do now to boost each component?

5 The reality of risk

In these days of economic uncertainty, how organisations respond to risk can not only affect how they fare in an evolving and fractious marketplace – but also how they come to be viewed by the public. As a result, many organisations are now putting significant resources into predicting and managing risks before they've even occurred.

Coming in myriad forms, risk should be thought of anything that prevents an organisation from meeting its objectives – and recognised as something that can seriously impact its operations, its people and the community in which it's located.

In this chapter, we'll explore themes around risk – including risk aversion, reputational risk, risk intelligence, risk analysis and, crucially, risk management – and we'll get a greater insight into why dealing with risks promptly and effectively is not only the cornerstone of safety work – but also the core of organisational success.

The idea of managing risk has obviously been around for a very long time. It's easy to imagine the caveman peering out from his rock home to check whether there's a Tyrannosaurus Rex wandering past his door. Despite this, risk management as an organisational function is still a relatively modern concept.

The Health & Safety Executive (HSE), Great Britain's watchdog for work-related health and safety, defines risk management as: "the eradication or minimalisation of the adverse effects of risks to which an organisation is exposed".

In their influential research paper 'Dealing with Risk: A Practical Approach' (1996), Professors of Operational Management, Fred Heemstra and Rob Kusters, describe it as the process of "identifying and responding to potential problems with sufficient time to avoid crisis situations".

When safety professionals engage with workers from different departments and ask about 'risk' or 'safety,' they're often met with an incredulous look and a shrug. This is routinely followed-up by the worker stating that it has nothing to do with their role within the organisation – and, to prove this, they'll explain what department they're in or what their job title is.

Surely, though, the words in your job title don't actually define your role, do they? Whilst there are obviously going to be safety systems in place to manage risk within the organisation, systems don't do this alone. Each and every worker should be aware of risks relevant to their work tasks and be able to prioritise actions to effectively manage these risks and report them through the management systems.

Risk has a major impact on organisations, and how they respond to the various risks the organisation encounters can influence how it operates as a whole. However, correctly assessing risks within organisations can be a considerable challenge – and not solely due to individual perceptions.

Classifying and assessing risk

In simple terms, we can consider three categories of risk – operational, strategic and compliance.

- **Operational** – anything that prevents the running of everyday operations;
- **Strategic** – anything that disrupts planning or organisational strategy;
- **Compliance** – anything that interferes with governance or regulatory obligations.

The wide-ranging and disparate nature of risk often makes evaluation daunting. Risks can span many areas – from natural to technological, to the physical to the psychological – and they are not always easily identified, never mind assessed. However, it is important to seek out and confront risk, as understanding, analysing and effectively managing them is often central to organisational success.

Risk can also be described as having three component parts – probability (the likeliness of an event occurring), the severity (of the particular event) and the exposure (the opportunity of it occurring). This is often expressed by the following formula:

Risk = Probability × Severity × Exposure

With these different parts – which all work independently – there are scales of tolerability. What's acceptable to one person may be intolerable to another. It's all a matter of perception.

Risk perception is typically a result of an individual's 'cognitive consideration' – thus, the way most individuals view and assess risk often depends entirely upon their own emotional subjectivity and personal experience. For example, an individual that has fallen off a ladder and injured himself is far more likely to view ladders as high risk than someone that has not. Humans are naturally biased due to the way the human mind works. Our memories are replete with 'reference points' anchored deep within our minds, which ultimately help to shape our opinions. In this instance, when the idea of 'ladder' is posited, the evoking of the bad memory is likely to surface much more quickly than the other much more mundane – and, thus, easily overlooked – moments in his life in which he has come into contact with them. Conversely, an individual that has worked on ladders every day for ten years – and never had an accident – may view them as entirely safe.

This can be dangerous. As well as giving rise to irrational decision-making, individual bias also leads to significant inconsistencies which can have potentially serious implications regarding the implementation of risk-management processes.

A strong safety culture consists of stated and shared perceptions in relation to safety and its management. This stems from the idea that the level of risk perceived in any given task can potentially alter the level of safety management applied to it. However, research shows that these perceptions often differ – often wildly – between individual workers. In his 2002 paper on the subject, Dennis Krallis shows that because perceptions of risk differ between different workers, they can often be rated extremely inconsistently as a consequence. If we return to the example of the man with the heightened sense of risk about ladders, whilst this view may be seen as some eccentricity on his part by his co-workers, he is also able to show statistical evidence that, due to his intervention within his area at work there have been no ladder-based accidents.

Risk aversion

The concept of 'risk aversion' refers to the behaviour individuals exhibit when confronted with situations of uncertainty. When confronted with a number of options in which the outcome is not clear, many individuals will naturally proceed down what they perceive to be the path of least resistance – however, a 'risk averse' individual will err on the side caution and always choose the option which they believe represents the safest choice.

It sounds like a sensible choice – and, indeed, it must surely be how most 'safety aware' organisations should operate, right? Well, before committing to this ostensibly prudent philosophy, it would be germane for business leaders to imagine what an organisation led by risk-averse people would actually be like.

Good safety professionals don't teach organisations to be risk averse – which, obviously, is not to say that they advise organisations to take risks. What they do is promote risk-awareness and ask organisational leaders to step up and embrace risk.

When organisations react to risk in ways not based on informed decisions, it usually constitutes an over-reaction – and this often leads them into either operating with over-confidence, or by becoming risk averse. Both can have disastrous results for organisations. The corporate attitude needs to move from risk aversion to a balanced risk-taking.

Reputational risk

Very often, the extreme reactions of organisations are as a result of the pressure felt by them around their reputations. 'Reputational risk' has become of major importance to organisations in recent years. Indeed, Harvard Business Review has long-heralded that in an economy where brand equity and intellectual capital are key, the biggest risk to organisations today is to risk their reputation (Eccles et al., 2007).

The world's media now regularly runs stories about multinational organisations and their supply chains – such as child labour being used in the production of clothing for fashion brands, poor safety performance in factories and other ethical issues like incidents of bribery or corruption.

Historically, multinationals have found it fairly easy to distance themselves from incidents such as these – which tend to take place in far-removed locations. However, as globalisation continues, and the commercial world continues to shrink, this is liable to change. Organisations that wish to safeguard their reputations need to consider all elements of their supply chains.

The supermarket giant, Tesco, faced a supply chain issue in January 2013, which went on to create a huge reputational risk issue – when it was discovered that their own-brand beef burgers actually contained around 30 per cent horse-meat.

The effect of this startling revelation was two-fold. As a result of the news coverage, consumers understandably lost confidence in the brand, resulting in a 1 per cent drop in the chilled and frozen food sales in the subsequent 13 weeks. Moreover, there was also an immediate response from the stock market, with Tesco share value dropping – also by 1 per cent – and wiping around £300 million off the value of the company practically overnight.

An organisation's reputation is a curious thing – and it can prove to be remarkably fragile. One famous case in the UK concerned a businessman called Gerald Ratner, who, whilst addressing a conference of the Institute of Directors at the

Royal Albert Hall in the early 1990s, made a speech which effectively destroyed his business. Ratner, who ran an extremely successful chain of high-street jewellers made a joke about the earrings sold in his shops being: "cheaper than an M&S prawn sandwich, but probably wouldn't last as long" – as well as other disparaging remarks about the shop's wares.

After the speech, which was reported heavily in the UK media, consumers began to avoid the shops and the value of the Ratner Group plummeted by around £500 million.

A holistic approach

Over the course of the last century, how organisations react to risk has progressed rapidly. Whilst a large part of this is due to the considerable research on the subject, the scope of various competing models and theories, and the adoption of new regulations, it is also in part down to the media coverage of the big disasters – such as Deepwater Horizon, Rana Plaza, the BP oil spill, and, more recently, the Grenfell Tower in London – and the public furores that grew up around them.

Whilst this has all contributed to a change in the corporate mindset and many commercial sectors adopting more considered approaches to risk, there is still significant opportunity for improvement. Business leaders and senior management teams need to take a much more holistic view of risk, and develop strategies that are integrated and business-wide.

To successfully embed risk management across an entire organisation, robust policies and procedures are essential, but, beyond that, organisations need to consider how best to circulate information about risk at every level because, at the end of the day, the most effective risk-management strategies are about communication and worker engagement.

Not only do workers need to understand what the risks are, but, also, they need to appreciate what can be done to report, mitigate and control them. This is what is known as Enterprise Risk Management (ERM): a holistic framework in which risks are viewed together, rather than having separate frameworks for different categorisations of risk. This is useful, because simplicity and clarity are central to effective risk management and control.

Organisations need efficient and effective monitoring capabilities, rooted in a robust, yet flexible, risk management framework. The three lines of defence describes how an organisation might ensure the risk is being managed correctly.

The best-in-class organisations take a three-pronged approach to monitoring risk:

1. Operational management/business leaders take ownership, responsibility and overall accountability for assessing, controlling and mitigating risks.
2. Operations are covered by several layers of internal governance – such as compliance, risk management and IT – and focus on supporting the execution of effective risk-management practices and assisting risk owners in reporting adequate risk-related information across the organisation.

3 Independent internal audits provide assurance to the operational management/ business leaders, which deliver detail on how effectively the organisation assesses and manages risk – including assurance on the effectiveness of the first two lines of defence.

Risk management is a continuous journey for every organisation. As we learned in the previous section, the solution is not to avoid risks, but understand them – and the best way to do that is to open up lines of communication around it. Effective dialogue, conversation and communication is every organisation's best line of defence against risk – only when every worker, at every level, is engaged and cognisant of potential risks will it be possible to manage them appropriately.

Aligning perspectives

In large organisations it can be challenging to create alignment on organisational risks. Yet without a collected view, and values that are shared globally, there can be no global power perspective – and since values are always rooted in culture, and culture is often based on sociological and religious values, creating a 'world view' seems impossible.

For example, when the League of Nations was created in 1919, following the end of the First World War, it was as an attempt to create peace through international accord. However, 20 years later the Second World War broke out and it was disbanded. Immediately following the carnage of the Second World War, in 1945, the United Nations (UN) was formed for the same reason – in an effort to ensure peace. But as this was enforced by the war's victors it immediately courted controversy, especially as initially some countries – notably France – were excluded from playing any meaningful part.

Though the UN has survived, it has endured decades of criticism both from its member states and those outside. During the 45 years of the Cold War between the USA and the USSR, both sides repeatedly accused the UN of favouring the other. The organisation's reputation was so bad in America in the 1970s that it even led to a pejorative bumper sticker sporting the legend: *You can't spell Communism without UN* becoming a bestseller.[1]

Over the years, critics have accused the UN of inaction, bureaucratic inefficiency, waste and corruption. This includes the organisation's famously timid response during the Somali civil war in the 1990s, which meant it missed opportunities to prevent significant human tragedy; and the Oil-for-Food scandal during the early 2000s, which saw Iraqi nationals trading oil with UN officials for food and other basic provisions to relieve the pressure of imposed sanctions.

The concept of Global Governance was introduced following the dissolution of the USSR, mainly in an effort to end the Cold War and bring the Eastern and Western blocks together. Its aim was to negotiate responses to problems that affect more than one state or region. However, underlying differences endured.

Americans and Eastern Europeans, for example, have historically different views on tolerance for inequalities and the growing demand for redistribution, attitudes toward risk, and over property and human rights continue to cause issues. In certain cases, searching for common ground in terms of outlook, serves only to accentuate differences.

In recent years, there have been many instances where the ideals of tolerance and free speech – as enshrined in the American constitution – have conflicted with the unswerving nature of Islamic Law. In the same way that the US's 'Right to Bear Arms' often causes conflict with other countries, especially following events like the mass shootings at Columbine, Las Vegas, Orlando and Texas.

Reasonable practicability

In the modern world, many people complain about the way health and safety works – they believe it has gone too far and that, rather than being restricted in what that can and can't do, they should be allowed to rely on their innate 'common sense' and many business leaders complain that they are unfairly pressured to pay out disproportionate damages.

Whereas, on the other side of the argument, many professional bodies and unions don't believe that legislation has gone far enough.

Legislation in most countries dictates that organisational risk should be controlled as far as is 'reasonably practicable' – but is this the case in practice?

Since it was first introduced, organisations in the UK have found the concept 'reasonably practicable' problematic – mainly due to the fact that it was never actually defined in the HSWA 1974.

The HSE's website attempts to qualify what is meant by the term by quoting a statement by Lord Justice Asquith during a 1949 Court of Appeal case – *Edwards v National Coal Board*. With use of some entertainingly inaccessible language, the judge attempted to shed some light on the ambiguous phrase by stating that:

> Reasonably practicable is a narrower term than physically possible and seems to me to imply that a computation must be made by the owner, in which the quantum of risk is placed on one scale and the sacrifice involved in the measures necessary for averting the risk (whether in money, time or trouble) is placed in the other; and if it be shown that there is a gross disproportion between them – the risk being insignificant in relation to the sacrifice – the Defendant discharges the onus on them.

The idea of 'reasonableness' is also a challenge as it's not obviously quantifiable – and, clearly, different situations will require different treatment. There is never a one-solution-fits-all in safety. What a business owner thinks is reasonable in terms of compensation following a worker's accident might be entirely at odds with what the worker thinks is reasonable.

Asquith's ruling was that risks *must* be averted – unless there is a *gross disproportion* between the costs and benefits of doing so. So, for example, whilst it's possible that a worker might fall off a chair in the staff canteen, it is not reasonable to expect an employer to remove all the chairs in order to prevent such an accident happening. Whilst the costs of doing so might be negligible, clearly the likelihood and seriousness of such an accident taking place needs to be weighed against the benefits of giving workers the chance to sit down and have a relaxing lunch.

Whereas, upon discovering asbestos in a workplace, an employer *must* remove it – even though it might be expensive to do so – because the risks of workers developing asbestosis, mesothelioma, asbestos-related lung cancer and other fatal diseases are so high.

Accept or tolerate

Another issue is how to deal with the risk – whether to accept or tolerate it? If risk is *accepted*, it's tantamount to simply conceding that it is a constant unavoidable. If, however, risk is *tolerated*, it is more like consenting to its existence – but with the caveat that something should be done to reduce it. It is obviously preferably to tolerate risk – and, thus, keep an open-mind to it and how it might be further decreased in future.

Any risk assessment process needs to be based upon tasks, and tolerability of risk needs to be determined from there. Only then will assessors be able to correctly measure whether the correct degree of control is being applied to the situation.

Risk assessors also need to look carefully at workers in order to assess their ability to perform certain tasks. In high-risk work environments, evaluating the capability of workers will naturally be of significant importance when considering the tolerability of risks. Fairly obviously, in high-risk environments, a new, or inexperienced worker will more likely constitute a greater risk than a seasoned old-hand.

The 'risk appetite' of an organisation may also affect what is considered tolerable risk. For example, theme parks – which typically cater to young audiences and parents – often need to be very risk-averse in order to maintain a solid public image.

Other organisations – such as those operating in the construction sector – inherently have much higher risk tolerance, mainly on account of being historically very dangerous. Indeed, within the UK, builders have the unfortunate distinction of suffering the highest number of at-work deaths.

Risk intelligence

Despite developing robust risk strategies, implementing ongoing monitoring and analysing progress reports, many organisations will experience significant and unexpected safety issues. Though there are potentially a number of reasons for this, frequently the cause of failures within risk management is the prevailing attitude towards risk mitigation within organisations.

Even organisations that are forward-facing and have an effective infrastructure in place will often struggle when it comes to fostering a sense of risk awareness, responsibility and, what is commonly referred to as, 'intelligence' throughout the fabric of an organisation – which can result in operational oversights and deficiencies.

Typically, when used in the context of safety, the term 'risk intelligence' is used to describe integrative processes related to governance, risk and compliance within organisations. It is usually considered to be where progressive organisations' cultures are when it has gone beyond the precepts of normal risk management.

Leo Tilman, Columbia University professor and leading authority on risk and corporate strategy, describes risk intelligence as:

> The organisational ability to think holistically about risk and uncertainty, speak a common risk language, and effectively use forward-looking risk concepts and tools in making better decisions, alleviating threats, capitalising on opportunities, and creating lasting value.

Risk intelligence demands that every individual in an organisation take responsibility for managing risks in the day-to-day operations. To cultivate a risk-intelligent culture, organisations must begin with a thorough risk assessment plan which details clear objectives, for workplace assessments, surveys and interviews; gap analysis, action prioritisation, implementation of risk control and the keeping track of the effectiveness of the approach. Comparing the existing approach against other influential factors such as governance, policies and procedures, competence, relationships, performance and accountability will help the top management understand the current state of play and the level of contribution of existing risk initiatives to create a positive impact on the business's intelligence of risk.

Strategic risk management

In an era where organisations are confronted with substantial increases in both competition and uncertainly, strategic planning and 'future-proofing' against risks has never been of greater importance.

Strategic risk management is the process of putting in place procedures and operations necessary to safeguard the future of an organisation. As a management system, it links strategic planning and decision-making with day-to-day operational management. It involves implementing measures to reduce the likelihood of losses occurring – and plans to ensure financial recovery in the event of losses, should they prove unavoidable.

Frigo and Anderson (2011) state that "Strategic risk management is focused on the most consequential and significant risks to shareholder value." If this is true, it could be argued that it is vital to the survival of an organisation.

Strategic risks are actions usually taken by organisations in order to gain some competitive or financial advantage, but which are liable to cause loss if unsuccessful. They are described by Allan and Beer (2006) as "threats or opportunities that materially affect the ability of an organisation to survive".

Failures may arise from poor decision-making, substandard implementation or changes in the business environment. A key part of the successful management of strategic risks is the ability to identify risks in a highly unpredictable environment. Key here is senior management implementing a holistic risk strategy to allow for the protection and support of the organisation's strategic aims and objectives, whilst supporting organisational sustainability. Strategic risk management also requires effective top-down lines of communication to impart a positive risk culture within the organisation – and its systems and processes.

Getting away from fear

Harvard University Professor and Risk Perception Consultant, David Ropeik, in his 2010 book entitled *How Risky Is it, Really? Why Our Fears Don't Always Match the Facts*, focuses on the psychology of how humans perceive risk. The central theme of Ropeik's book, which owes a good deal to the research of psychologist Paul Slovic, concerns the human 'perception gap' – how people continue to engage in risky behaviour, whilst simultaneously worrying unnecessarily about other things of little consequence.

I've always been intrigued by a friend's paralysing fear of spiders. Whenever she spots one in her house, she will usually become rooted to the spot, then her arm will slowly rise, and a quivering finger will point across to it. This is typically followed by me grumbling that it's 'just a spider', before traipsing off to find a cup and a newspaper in order to remove the wayward arachnid. Mosquitos, however – which are far more likely to cause her actual harm – she takes little issue with, literally brushing them off without a second thought.

As a safety professional, quite often, when people find out that one of my hobbies is swimming with sharks, they seem surprised. However, sharks get a totally undeserved bad press, likely due to *that* movie back in the '70s. Indeed, *The Shark Attack File*, maintained at Mote Marine Laboratory in Sarasota, Florida, contains statistical information about local shark attacks. Since the year 1580 (it goes back that far!), around 2,200 attacks by sharks on humans have been logged. To put that into context, it's approximately 5per cent of the number of Americans injured by toilets each year.

Whilst our perceptions of risk – and associated fears – are obviously personal issues, largely framed by our experiences, how we deal with them with can be remarkably different. Ropeik suggests that individuals are all different places on the 'risk continuum' – a scale with 'risk averse' (those who attempt to avoid risks at any cost) at one end and 'risk prone' (those who blithely accept risks) at the other.

Whilst is difficult to change an individual's place on the risk continuum, an awareness of the psychological factors at work when faced with a risk, can help them to achieve a balance between being risk prone and risk averse. This may help them make more considered decisions and avoid fear.

In an excellent article for *Harvard Business Review*, 'Understanding factors of risk perception', Ropeik identified 14 factors that affect risk perception. Whilst it should be pointed out that some of the factors may be in conflict with others, or have a stronger effect than others, let's have a look at them here.

Trust versus lack of trust: Trust has a huge effect on how individuals perceive risks. The less they trust the people informing them about a risk, the people/organisations supposedly protecting them from it or the processes determining their exposure to a risk, the more afraid they become.

Imposed versus voluntary: Individuals are typically more afraid of risks that are imposed on them than when we voluntarily expose ourselves to the same risk. Which perhaps explains why erratic driving is always more of a worry for passengers, than drivers.

Natural versus man-made: If a risk is natural, such as adverse weather conditions, individuals are less afraid than if it's something deemed artificial or man-made. This may explain excessive public fear of pesticides and industrial chemicals.

Catastrophic versus chronic: People are naturally more afraid of things that can kill *en masse*, suddenly and violently and all in one place – such as a bomb – over things like heart disease, which, though causing hundreds of thousands more deaths, does so slowly, individually, and not all in the same place.

The dread factor: The worse the outcome from a risk, such as being savaged by wild animals or being boiled alive, the more afraid of it we are.

Hard to understand: The harder a potential risk is to understand—such as nuclear power or industrial chemicals—the more afraid individuals are likely to be. When a risk is invisible – like a gas leak – the fear becomes more pronounced.

Uncertainty: This is less a matter of the science being hard to understand and more a matter of not having enough answers. When HIV and AIDS were first clinically observed in the USA in the early 1980s, there was widespread fear about it. Now that the virus is better understood – and media moral panics have abated – public fears have dropped massively.

Familiar versus new: When risks are first encountered, individuals are much more afraid than after they have lived with the risk for a while. When mad cow disease first showed up in just a handful of cows in Germany in 2000, a poll found that 85 per cent of the public thought this new risk was a serious threat to public health. That poll also asked people in Great Britain, where the disease had killed more than 100 people and hundreds of thousands of animals. But because people in the UK had lived with it for 14 years, only 40 per cent of them said mad cow disease was a serious threat to public health.

Awareness: When the news is full of stories about a given risk, like ozone depletion, our fear of that risk is greater. For example, on July 3, 2002, amid a flurry of "Will the terrorists strike on July 4th?" stories, the FBI said requests for handgun purchases were one-third higher than expected. But awareness doesn't just come from the news media. When, as individuals, we've recently experienced something bad, such as the death of a friend or relative to cancer, or witnessed a crime or an accident, awareness of that risk is greater, and so is our fear.

A known victim: A risk that is made real by a specific victim, such as the recent child abductions making news, becomes more frightening, even though the actual risk may be no greater than it was before it was personified by this victim.

Future generations: When kids are at risk, our fear is greater. Asbestos in a workplace doesn't frighten us as much as asbestos in schools.

Does it affect me? We don't perceive risk to 'them', to society, as fearfully as we do risks to ourselves. This explains the desire for zero risk. A person doesn't care if the risk of cancer from pesticides on food is one in a million, if he or she could be that one.

Risk versus benefit: The more we perceive a benefit from a potentially hazardous agent or process or activity, such as drugs or vaccines or skiing or bungee jumping, the less fearful we are of the risk.

Control versus no control: If a person feels as though he or she can control the outcome of a hazard, that individual is less likely to be afraid. Driving is one obvious example, as is riding a bike and not wearing a helmet. Control can either be physical (driving the car, operating the bicycle) or a feeling of controlling a process, as when a person participates, setting risk management policy through involvement in public hearings or voting.

It's important to keep in mind that fear is an emotion that can't always be calmed by statistics and facts, so people do not always act rationally. Decision making including risk assessment is a complex process affected by participants' individual beliefs and life experiences.

Awareness of your and others' perception of risk is the first step to more effective risk assessment and management. Our perceptions shape our decisions, so it can be useful to explore these factors while making risk decisions.

Imagination and risk

A large part of risk management is about being able to visualise the unexpected – and, for this reason, a degree of abstract thought is required.

The imagination operates within the non-rational part of the mind – and, as such, in most respects, it is of limited use in the safety industry. After all, within organisations with procedural safety systems, governed with heavy legislation, there is usually little room for creative thinking. Indeed, the largely bureaucratic mind-sets that inhabit the domain of workplace safety can often frustrate creative people.

There is a danger, however, when taking a predominantly rationalist approach to risk management of the practitioner, and the organisation settling too much into a comfort zone.

The more we attempt to systematise the management of risk, enforce policies and procedures and fixate on the legal and regulatory aspects, the greater danger there is of them actually losing sight of what it really is. Risk management is not simply a series of box-ticking exercises, it requires application, thought and vision. It needs to return to being a person-centred process, in which systems serve specialists rather than the other way around.

Instead of developing even elaborate and complex systems, safety professionals may do well to focus less on compliance and simply set aside some time each week to look around themselves and try to imagine the possibilities.

THE NAKED TRUTH: THE REALITY OF RISK

To implement a robust risk management strategy across an entire organisation, strong policies and procedures are essential, but so too is the need for leaders to look beyond formal frameworks and reflect on how they understand and communicate about risk.

Whilst it's true that risk assessments, matrices and registers are good support tools, the most effective risk management systems depend on every member of an organisation's workforce understanding the risks associated with their own role – and being motivated to do everything they can to report, mitigate and control those risks.

It is only when risk management is integrated at every level that safety stops being thought of as just another useful 'bolt-on' service – and becomes, instead, an integral part of an organisation's decision-making process.

GET NAKED: THE REALITY OF RISK

1 How is risk is defined in your organisation? How have you arrived at this definition? Is the definition assumed or openly stated?
2 What is your organisation's attitude to risk? Where would you consider it to be on the risk prone – risk averse continuum? How does this perspective manifest in the workplace?

3 What are the main operational, strategic and compliance risks to your organisation? How are these widely understood by everyone and discussed at high level meetings?
4 Do you consider safety to be included in an integrated approach to risk management within your organisation? If not, what could be done to align it? What could you do – personally – to promote this?
5 Ropeik introduces 14 factors that affect risk perception. Which of these could be playing a major influence in the way your organisation and its workers perceive risk?
6 How would you describe your organisation's appetite for risk? How does this affect everyday performance?
7 Choose a workplace task or activity that you know reasonably well. Take time to review the task from a risk perspective but without any formal structured approach. Allow your mind to imagine what risks there could be, and get creative on how these risks could be brought intelligently to a level of tolerability.

Note

1 I'm not sure they really thought this through – you also can't spell United States without UN.

6 The value of safety

Though some leaders now understand the vital importance of good safety practice to support business objectives, in the current period of pronounced economic uncertainty, many Occupational Safety and Health (OSH) practitioners are still faced with the unhappy prospect of trying to prove the value of safety to their organisations.

In this chapter, we'll explore the contemporary focus on attaining zero injuries, discuss why a more progressive approach to 'creating safety' would be more successful, look at the hidden costs of poor health and safety – and illustrate, unequivocally, why good safety is good for business.

Within the safety sphere, there is always much talk about 'zero' – often delivered in a hushed and slightly reverential tone.

For safety practitioners, there is something beguiling about zero. It seems on the face of it something reasonable, manageable, even attainable. But, in practice, it is more like trying to remember the details of some already half-forgotten dream. Though you might chase it, it remains elusive, obscure and teasingly out of reach.

The history of zero

Like everything else about it, the history of 'zero' is hard to pin down.

It seems like its existence – or lack of it – was arrived at entirely independently by both the Indians and Mayans, at around the third century.

In India, where they were already using the decimal system, they had been using an empty space to describe the concept of 'zero', but found this to be confusing, since they also used spaces to separate numbers, and so a 'zero' symbol was created to help describe nothingness.

Meanwhile, at exactly the same time, some 9,500 miles away, the Mayan civilisation of Mexico started using a zero in their calendars.

The Romans and Greeks, to whom we owe most of our mathematical learnings, did not use the number zero, since they did their calculations on an abacus. So, it was only in the comparatively late eighth century, that the concept of a numeral zero made it to Western civilisation.

Striving for nothingness

With OSH, the emphasis is always on zero accidents – and, to be fair, it makes sense – after all, what other goal could there really be?

The whole point of OSH is to create a working environment free from injuries and accidents, isn't it? So, a goal of 'zero' is the obvious target.

It's only when you think about it, that you realise that it's a fairly unusual goal. Indeed, striving for zero sets safety practitioners apart from other professional people – and means that they are habitually going against the grain. After all, in what other industry do practitioners actively try to achieve . . . well, *nothing*?

Being constantly focused on zero is additionally tough, because – as human beings – we are conditioned to think of zero as simply an absence of value.

In practically every other industry (just as in any other part of life), whether it be business, sport, personal development, finance or education, zero is akin to failure. For this reason, setting goals of zero happens rarely.

Even the language of 'zero' is rather unsettling. The word itself is derived from the Italian 'zefiro', likely a corruption of the Arabic 'safira' – meaning 'empty' or 'nothingness'.

As humans, we naturally see progress as gaining *something*, rather than achieving nothing, so at a psychological level, the whole framing process in terms of reaching zero is actually rather demotivating.

At a conceptual level, there is another problem. Zero is an absolute – and this makes humans uncomfortable. It seems at once somehow both too intolerant and too severe. When it comes to injuries and accidents in organisations the 'zero tolerance' and 'zero harm' demanded, would be inconceivable in other areas of a business. For example, it hardly seems imaginable that, within those same organisations, leaders would instil a similarly hard-line stance on mistakes made by managers?

In *For the Love of Zero: Human Fallibility and Risk* (2012), social psychologist Dr Robert Long states that: "The language of zero only inspires perfect people. The rest of us are motivated by patience, tolerance, understanding and the scope to learn and mature." Long suggests that the real outcome from an organisational focus on zero is usually just annoyance and confusion. When workplaces are beset by 'safety evangelists' in hi-vis tabards insisting workers down tools and fill out reports every time someone spills a coffee or gets a paper cut, it may look positive on paper – however, a quick scan of the organisation, and inevitably you'll find this sort of overzealous behaviour is met by deeply-felt frustration and a good deal of cynicism.

But what other choice is there? After all, zero makes complete sense, doesn't it? And, not only that, but it's easy to understand. People get it.

Well, that might be true – but, using zero as a goal or target is quite palpably absurd. However, that is not to say that safety professionals should simply accept that there will be accidents – and move on. No, there is no room for that kind of *binary opposition* when it comes to safety.

Binary opposition

Binary opposition is when something can be explained using two mutually opposing terms. Think of a radio – it's either on or it's off. If it's off, it's certainly not on. And vice versa. The central tenet of binary opposition is simple: if it's not one thing, it must be the other. There is no grey area, no 'in between' – no third way.

So, binary oppositionists think that if you don't believe in God, you must therefore believe in the Devil. If something is not 'white', then it follows that it must be 'black'. That if you don't endorse a goal of zero, then it follows that you must be endorsing harm.

Clearly, only a very peculiar and perverse safety professional would actively encourage injuries – so where does that leave us?

Whilst a goal of zero accidents may be the logical choice, the problem is that when an organisation dedicates the entire focus of their safety activities to attaining zero, it is usually followed by a number of quite negative consequences . . .

For example, as the management teams focus their efforts on attaining the elusive zero, it builds a culture of intolerance and increases workers' fear of failure. This can cause considerable stress throughout the workplace, and shifts the focus away from effective risk management to a much more pedantic area of risk elimination.

As it treats all risks – large *and* small – in exactly the same way, it naturally diverts attention from the real issues, and puts too much emphasis on the past, rather than looking at managing risks in the future.

Moreover, as these kinds of initiatives tend to be backed up by the threat of punitive measures, they drive accurate reporting underground and stifle honesty, clarity and conviction. In some extreme cases, they even foster a climate of deceit.

Obviously, all people involved in workplace safety want to make the organisations they work for be as safe as possible – and, indeed, strive for it every day. However, spending all your time chasing zero simply becomes a distraction to the job in hand.

The dedicated focus on zero as the goal is risky, because it risks:

- instilling a culture of intolerance;
- focusing on failure;
- increasing fear of failure;
- distorting our data;
- driving down reporting;
- diverting attention away from the major issues to focus on all risks;
- crushing creativity;
- engendering a reactive mind-set;
- stifling openness, transparency and trust;
- killing flexibility;
- reflecting the past, rather than the future;
- promoting penalty and punishment;
- raising worker stress levels.

Adding value

So, let's forget zero for a minute – and think about how OSH might go about adding value instead. When it comes to safety performance the emphasis should always be on measuring specific safety objectives – not the number of accidents that have (or have not) occurred. Do it correctly, and the creation of a workplace free from accident, injury and ill-health will be the outcome, rather than the goal.

Instead of concentrating on bringing down numbers, organisations need to look towards the future rather than dwell in the past and, I suggest, switch from focusing on preventing accidents to looking forward, at the inputs that can help us in creating safety.

When accidents or near misses take place, it is vital to use them as something to be learned from. Rather than just recording the root causes and moving on, these should be utilised as the starting point of a discussion to create a shared-learning experience for workers. Putting numbers onto charts is not a viable defence against accidents – but safety conversations, meaningful training programmes and responding to employee feedback can be.

Achieving high levels of safety performance needs more than just diligence or devotion to duty, it requires means and methods. Look at what positive contributions are being made to improve safety at organisations every day – the team talks, shared learning experiences, Lean Six Sigma projects, worker engagement, safety conversations, the impact of training and worker suggestions and feedback loops. There's so much to consider it shouldn't be hard finding things in your organisation to focus on. Consider how you can measure the impact of these activities and initiatives. Organisations need to devise ways to measure these input activities, use them to create and implement best practices – and, when this has occurred, use this data to recognise and reward brilliant contributions and leverage improvements. Wouldn't you agree that focusing on something important and tangible is a far better idea than chasing nothingness?

Cutting costs

Since the global economic downturn hit, health and safety roles have been relatively thin on the ground. However, the Organisation for Economic Cooperation and Development (OECD), a body that represents the 30 most industrialised countries in the world, has recently forecast an imminent economic recovery. After years of fairly dispiriting conversations with health and safety professionals mainly centring around resource constraints, downsizing and fighting for survival, it looks like boom is hopefully about to follow bust.

What do safety practitioners do when in times of economic difficulty? Well, from the conversations I have been having over the last few years, it seems that a good many of them do as little as possible. That's not to say that they've been idle, but as survival instincts kicked in, many clearly decided that the best course of action was to keep their heads down and get on with the work; keen not to draw attention to themselves or rock the boat any more than was necessary.

Whilst it's true that during periods of economic uncertainty, leaders are likely to be preoccupied in other areas there are always ways in which safety teams can add more value.

These might include:

- Freeing up finances by investing in more effective assurance approaches;
- Reducing the focus on conventional auditing;
- Reviewing operations and revising or eliminating those that represent unacceptable risk levels.

The most important part of any organisational leaders' role during a period of economic uncertainty is trying to manage that uncertainty. This is often problematic because it is never clear when the dark days may end – and a bright new economic dawn may be upon them.

Not unreasonably, many leaders struggle to make far-reaching decisions about how much effort their organisations should place on survival, and how much they need to spend on positioning themselves in the ensuing period of recovery.

Rather than attempting to simply ride out the bad periods, progressive safety practitioners will keep adapting – and keep maintaining the flexibility of their programmes. In doing so, this will aid them in avoiding the need to make such difficult decisions themselves, whilst reminding leaders that good safety is good business.

Good safety is good business

The mere mention of 'health and safety' is often enough to strike fear into business leaders. Indeed, I've seen General Managers and CEOs recoil in shock or judder to a complete standstill when the subject is raised. However, with increasing frequency, what I tend to hear most is leaders complaining wildly about ever-increasing health and safety legislation and growing levels of bureaucracy.

In fact, the amount of bureaucracy is actually falling. Even the EU – which often garners a bad press for being overly-administrative, fussy and officious – is currently going through the exhaustive process of reviewing and rewriting its safety legislation and attempting to condense and simplify things.

Whilst it's true that successive US administrations have basically ignored health and safety, and recent UK governments have actually gone further by actively coming out in opposition to it,[1] it's quite odd that there's probably more written on the subject of health and safety now than ever before.

In part, this is no doubt due to the culture of 'ambulance chasing' that has taken root in recent years – those civil cases in which members of the public are encouraged to make safety claims on a 'no win, no fee' basis.

Whether these civil claims are successful or not, they feed into the popular narrative about 'the nanny state,' and cost governments and industry hugely – and not just financially, but also in terms of time and stress for the parties involved.

For anyone that runs an organisation, preventing workers from being harmed and providing a safe environment for them to work in is not just a nice thing to do – it is a legal requirement. But much more than that, the price of getting it wrong can be huge – such things as bad publicity, criminal prosecution, civil claims and loss of the license to operate might very well do significant damage.

On top of this, losing skilled workers – even if it's just for a few days – can have serious consequences. In some of the smaller organisations that we consult with, these businesses often have extremely stretched resources and little in the line of contingency planning when it comes to accidental losses. Though they will naturally hope for the best, it's true that a serious accident can certainly put smaller organisations quite literally out of business.

Ask most business leaders what they will do in the event of a serious accident and they'll look pensive for a moment and then exclaim: "well, it's covered by insurance!"

Most employers think this, but it's rarely that straightforward. Insurance policies are usually very specific about what they cover. They may pay out for serious injuries or damage, they may not. And, of course, there's the excess still to pay.

Whatever is met by the insurance company, there will always be a list of other costs that will have to be met by the organisation.

The costs that grow up around accidents and injuries can typically include:

- lost time;
- worker cover/extra wages;
- overtime working;
- sick pay;
- damage or loss of product;
- equipment repairs;
- workplace maintenance;
- production delays;
- accident investigation costs;
- fines;
- structural improvements;
- legal costs;
- lack of business due to loss of reputation.

Despite the withering responses that matters of health and safety often earn from business leaders, it is a subject that deserves significant thought – there are just too many factors that might have a detrimental effect on your organisation if they are not considered fully.

However, if it is to be assumed that – aside from all the bureaucracy that they seem to dislike – most business leaders are still keen to work in organisations where their co-workers are not killed and injured every day, where they are not constantly paying out money on fines, financing absent workers (whilst also paying for replacements) and, at the same time, enjoying a terrible reputation in their own industry, they really need to strive for better.

Hopes and aspirations are always difficult to quantify – and attaching a number to them doesn't make the process any easier. So I suggest that organisations need to forget the ideology of zero. It's simply nothing more than a distraction from what the real focus of what OSH should be – creating a workplace that is populated by happy, healthy workers who are engaged, committed and looking out for one another. When this occurs, a wave of positive energy streams through the organisation and productivity naturally soars. This is the real value of safety.

THE NAKED TRUTH: THE VALUE OF SAFETY

The focus on reducing worker harm to as low as possible is an established principle of organisational health and safety management but it should not be static.

Instead of chasing zero injuries, which is always something of a distraction and regularly culminates in safety practitioners losing sight of what they're really supposed to be doing, the focus needs to be on creating safety in the workplace by focusing on doing more of what keeps people free from harm rather than trying to stop things going wrong and causing injuries.

In this way, it is possible to work in a more positive fashion – but also in a way that doesn't impose a culture of intolerance, increase fear or drive reporting underground. Good practice in health and safety is not only a legal right for workers – but also makes sound business sense.

When organisations create a positive, safe work environment they reduce absences and sick leave. Workers are more motivated and staff turnover falls. It can also mean that the organisation maintains and even improves its market reputation.

As well as the boost in productivity and profits that occurs naturally in organisations when workers are happy and healthy, there are additional bottom-line benefits including lower insurance premiums, and, crucially, that the organisation retains its license to operate.

GET NAKED: THE VALUE OF SAFETY

1 What target does your organisation set for safety? What are your thoughts on this? Is it an appropriate target that is both meaningful and achievable?
2 Do you think that your organisation promotes a culture of learning and tolerance, or one of absolutes and reactivity?
3 Consider a 'good day' at work in your organisation. What are the things that contribute to success? Perhaps it's solid direction from leadership, effective management, good teamwork, clear communication – what else? Now, which of these things do you believe there needs to be a focus on in order to *create safety*?
4 How could you provide a *value gain* to motivate people in safety?
5 Instead of chasing zero, the concept of creating safety encourages us to strive for safety excellence. What things might you be able to achieve in your organisation that would reduce corporate costs but maintain safety value whilst delivering excellent performance? Where will you begin?

Note

1 Former UK Prime Minister David Cameron famously said in 2012 that he was planning to kill off the 'farcical and marginal' safety culture – to cut back the red tape and allow businesses to flourish.

Part 2

Getting down to business

7 Corporate culture

Corporate culture is the product of an organisation's values and behaviours collectively shaping its distinct social and psychological environment. It represents the shared beliefs and principles of the workers, and dictates the norms, assumptions, beliefs and habits that exist.

Often considered as 'the way we do things around here,' corporate culture is frequently implied, rather than expressly defined, and – if not given specific direction – will develop organically over time.

This chapter provides a primer to our thinking about culture. We'll look at how culture plays a vital role in an organisation's success – and how cultivating a positive workplace culture can maximise efficiency, sustainability and performance.

Corporate culture is typically and simply defined as 'the way things are done around here'. Essentially, it's a resulting ethos and set of customs – often as a sort of by-product – from the prevalent attitudes, beliefs and values of an organisation, and shared by the workers.

The idea was popularised following the publication of the bestselling book *In Search of Excellence* (1982) by Tom Peters and Robert Waterman, in which the two business strategists argued that operational success can often be solely attributed to an organisation's culture – so long as it's positive, empowering and people-oriented. You can imagine the results from a negative, disempowering and selfish organisational culture.

Peters and Waterman argue that the culture of an organisation is largely invisible to individuals inside it – in much the same way that water is to a fish swimming in it. Even though the water affects every aspect of the fish's thinking, its behavioural patterns and its social interactions, it is almost entirely unaware of it.

Examples of the manifestation of corporate culture can be found in branding and colour schemes, the organisation's mission statement, company policies, incentives and rewards for workers, training and development and much more. Yet, unless considered critically, much of this surrounds us invisibly like the water around the fish.

Tom Davenport, Academic Director of Process Management Research at Babson College in the United States of America, sums it up rather succinctly: "Think of culture as the DNA of an organisation – invisible to the naked eye, but critical in shaping the character of the workplace" (Neilson and Pasternack, 2005).

Though corporate culture can be difficult to observe from within the organisation, it can be vital to operational success. When organisations are failing to perform, leaders often decide they must make changes to the corporate culture – recognising the fact that it's as important as more prosaic factors, such as technological advance or organisational structure.

Dutch social psychologist Geert Hofstede (2002) describes culture as a set of unwritten rules of behaviour that set out what a particular group expects its members to do – and what to believe. This makes it much more powerful in controlling and managing behaviour than any rules and regulations.

For Hofstede, culture is defined by the 'collective programming of the mind within a group' – and, as such, understanding an organisation's culture is intrinsically important to its 'success'.

If values and behaviours of workers are aligned with the objectives of the organisation, it will lead to competitive advantage. Whereas, where there is misalignment, it is more likely to result in performance problems and organisational failure.

In *Managing the Risks of Organisational Accidents* (1997), James Reason highlights the power of culture by citing an example from the world of aviation. He states that, on the face of it, the operational success or failings seen by commercial airlines should be fairly consistent. After all, they use the same aeroplanes, operate within largely the same environmental conditions, use the same air-traffic control system, and their crews function to similar standards. However, statistics showing the number of passenger fatalities occurring, presumably a fairly pertinent performance indicator, deviate wildly between companies – from 1 in 260,000 to 1 in 11 million. Reason thought that whilst resources may be partly responsible, most of the differential was due to cultural variations.

In organisations with strong cultures, workers will routinely express the sense that they 'belong,' which naturally increases their sense of loyalty and slows staff turnover rates. Workers also feel valued, which promotes high morale and increases motivation – which, in turn, inspires outstanding performance.

Time and again, a strong corporate culture has proven to be more important to organisational success than corporate strategy, market share or technological advantage. Understanding how to instil and manage culture can help business leaders to make a mark on organisations that will resonate far into the future.

In his book *Who Says Elephants Can't Dance?* (2013), former Chairman and CEO of American technology giant IBM, Lou Gerstner, gives an account of how he led the company back from the brink of bankruptcy to unprecedented business success "Until I came to IBM, I probably would have told you that culture was just one among several important elements in any organisation's makeup and success – along with vision, strategy, marketing, financials, and the like."

Gerstner concludes: "I came to see that culture isn't just one aspect of the game; it is the game. In the end, an organisation is nothing more than the collective capacity of its people to create value."

The complexity of culture

Corporate culture is complex and is made up of a network of inter-connected elements.

Rather than a hierarchical structure, these elements are intricately linked together and connected in a much finer, more subtle way, just like the silk of the spider's web.

As human beings, just like spiders on a web, we appreciate balance. Whether acrobats or tightrope walkers, plate spinners or ball jugglers, machine operators, risk management professionals or business leaders, feeling steady and balanced is good. Beyond having our feet on the ground and knowing what's what, we're constantly seeking a sense of psychological – as well as physiological – equilibrium.

Culture helps us achieve that by providing an invisible framework or 'web' upon which our emotional and cognitive states can be stabilised against the constant daily flow of external and internal stimuli that hold the potential to upset our balance. Just like the spider, whose network of fine silken threads provides the strength to support the creature, even when the wind blows or the rain falls.

REFLECTION POINT

In 2008, the global financial crisis struck. As many leading financial organisations found themselves dashed against the rocks, the key factor connecting each one was their insatiable desire to lead through performance.

Rather than deep, positive organisational cultures, the banks had become reliant on individuals for their success. Driven by achievement of goals and attainment of results, news reports explained that these organisations had developed cultures which were heavily 'process-oriented', effectively ignoring the collective contribution of the broader workforce.

In the wake of the crisis, we see these same organisations trying to invert their approach, and now being led by building 'values-based cultures' – rather than being driven only by performance strategy.

Despite this progress, it may be many years before these banks lose the negative stigma and rebuild trust.

The strength of the web is as vitally important to our organisation as it is to the spider. But this framework, comprising structural elements (terms of reference, rules, management systems and organisational hierarchies), cognitive aspects (attitudes, interpretations and meanings, values, beliefs, assumptions and personalities) and behavioural experiences (language, stories, norms, customs and practices) appears so complex and tightly interwoven that we may be deterred from attempting to tackle it.

Get organisational culture right, and it helps us avoid uncertainty, unpredictability and confusion, clearing the way to facilitate the pursuit of meaningful corporate objectives with a shared sense of strength, ownership, unity and harmony.

The power of positive culture cannot be overstated. Its effect is akin to the sun, impossible to see those individual rays of sunshine beaming down on us from above, but when you feel them on your skin the immediate sense of comfort, optimism, internal strength and ability quickly develops within and radiates from us.

But, like an untended garden quickly becoming overgrown with weeds, ignore that 'invisible framework' of culture at your peril, and you'll find the sunshine sliding behind the dark clouds as the stinging nettles, tumbleweed and poison ivy soon creep in to stultify learning, dent motivation, suppress innovation and draw a sharp halt to continuous improvement.

It's all too easy to blame culture for rocky patches in business, uncomfortable mergers or acquisitions, failed strategies or leadership misfits. Though, if we're not really sure what culture is, why are we so quick to lay the blame for failure at its feet?

Culture – an indefinable concept?

So, what really *is* culture?

We can understand that it's as complex and as delicate as a spider's web. We can see that it can have major impact on the success (or failure) of an organisation. Though, in our organisations, we tend to overcomplicate it.

We use phrases like 'performance culture'; 'people-centred culture' and 'winning culture' to find ways of describing our businesses, hoping that they draw attention, offer meaning, add value or differentiate our organisation, department or team from others. We'll say we have a 'mature culture' when we talk of the professionalism and gravitas we command, or the 'interconnected culture' that allows us to get things done. We set visions for the future as we try to attain a destination we refer to as a 'safety culture', whilst beyond our walls the media tells us how as a society we've moved from being a 'tolerant culture' and we now live in a 'culture of fear'.

But all we are doing is affixing labels. Providing pigeon holes. Succumbing to our inherent human instinct to enforce order on things. These labels don't help us. They tell us nothing until we begin to *describe* what it is that we mean. As ground-breaking industrial psychologist Leandro Herrero maintains, there is no entity called 'culture': it's just *ways* of doing things (Herrero, 2008). It's our beliefs and values that shape *what* we do, the norms and rules that govern *how* we do what we do and the attitudes that influence the *way* we do it.

Only if we are brave enough to dig deeper do we really see what culture is. It's just all about *behaviours*. When we describe culture, we describe the things people do, or don't do. We describe behaviours. Acceptable behaviours, or unacceptable behaviours.

Culture is a magical, mysterious, force. Invisible, in its own right, we only really see it when we piece together the behaviours like a jigsaw puzzle.

The personality of culture

Broadening the use of metaphor in this chapter, we can view culture as the grouped 'personality' of the organisation. Indeed, former Chairman of the Cadbury-Schweppes group, Sir Adrian Cadbury, once remarked that "the character of the company is in our hands. We have inherited its reputation and standing, and it is for us to advance them".

Let's take a step back for a moment to set some context around this idea of culture as character or personality. If you have a young child, think about how they learn their attitudes towards certain things or the correct behaviours to engage in an activity. If you don't, think back to your own childhood. Recall how you learned not to touch hot pans on the stove, how not to get too close to the fire, how to cross the road safely, how to tell the time, how to drive a car, etc.

Each of these behaviours we learned from people in positions of authority, or 'life leaders' if you will – whether parents, carers or school-teachers. We learn skills, mental perspectives (attitudes) and acceptable behaviours from those who strive to teach and lead us towards what is right or expected of us. With practice, we learn to abstract these attitudes and generalise the taught or instilled behaviours, and they shape the way we act and react, our everyday actions, and, ultimately, who we are.

As the years pass, the collection of attitudes and behaviours shape and form our own distinctive personalities, but think carefully and you may just recall comments from those closest to you that suggest that you are 'just like your father', or approach things in the same way your mother would.

In the work environment, it's exactly the same (though we may at times ignore or choose to deny it), except that our loving parents, guardians and favourite teachers have been replaced by managers and business leaders.

In some organisations, the personality of the leader may be so strongly influential that it shapes not just the behaviours of their direct reports, but, also in time, the personality traits passed down to managers influence those beneath them too. As we acquire and collect knowledge to interpret our experiences in the workplace, our behaviours become reinforced and if they join the similar behaviour patterns of those in our vicinity, they become the 'group personality'.

By natural extension of the 'group personality' point, when these are further endorsed and supported by the predominant behaviours displayed by management and peer groups, they develop into even broader collective behaviours and responses. These behaviours become commonplace, and evolve into norms, customs, rituals or habitual practices. And so, we have the shared humanistic approach, or 'organisational personality'. Or, as we may choose to call it: culture.

The strongest and most productive organisational cultures are those that do not depend on one individual, but instead are constructed through the personalities and behaviours of those people within them.

In these 'human webs' of significance each individual person contributes to the culture every day, though arguably with increasing influence as their authority and visibility increases within the organisation.

These 'relationship-oriented' cultures place great value on human elements such as teamwork, morale, engagement and communication – and it's here that organisations thrive.

When the culture is co-created and becomes so robust that it stands alone, without the necessity of individual stewardship by the CEO, it becomes self-reinforcing through inspiring workers to conform to it – without the need for prescriptive direction or the strength of an individual personality. It becomes 'the way we do things around here'.

To achieve this state, the corporate culture needs clear values – including the age-old elements of trust, honesty and transparency – between the organisation and its stakeholders, whether society, shareholders or the workforce.

Whilst in one sense culture can make the success of the corporation, at the same time it holds the potential to break it too. Culture can be a 'hidden risk' within our organisations. Ignore it, and it will soon catch up on us – snarling and biting like a rabid dog. Whether we want it or not, whether we like it or not, whether we manage it or not, culture will still exist within our organisation.

There's nothing we can do to shake it off. Every company, everywhere, has a culture that is unique. Your job as a leader is to ensure that it's a positive one.

So, beyond just *having* a set of values, leaders must elevate and discuss the values – and the organisation's goals and objectives – constantly and consistently: corporate culture is revised every hour of every day, with every thought, word and deed.

Hallmarks of culture

So, if culture is the way of corporate life that provides the framework for stability, learning, growth, success and employee satisfaction, it makes good sense to want to ensure a robust and positive culture within the organisation.

But, as we have seen, culture can mean different things to different people, and its dynamic, fluid nature means that it's challenging to define. So, what *can* we say about it? Well, as clear and agreed definition has proved elusive for the academic and scientific communities, we can choose an alternate route.

Let's look instead to discovering which elements stand out as identifiers or 'hallmarks' of culture:

- **A shared and common language utilised by members of the group or organisation**. Irrespective of the actual language choice used to conduct our daily business, each organisation (and in large organisations, perhaps even within each business unit or department structure) has its own *lingua franca* that helps it to communicate internally. Whether we consider it specialist terminology, technical jargon, organisational abbreviation or industry vernacular, listen carefully and you'll hear it. The words and phrases that, used so often, pockmark and gilt-edge your vocabulary.
- **Accepted social norms, customs and practices**. Think about the softer side of doing business in your business. For example, how do meetings commence? With allocation of a formal chair and minute-taker – or with ten or fifteen minutes of random chat amongst members about holidays, the weather and the new coffee capsules at the break-out area as they drift into the room? What's the first subject the CEO speaks about when he calls the team (or organisation) together for an update? What's the process for shift handovers on the production line? What do these accepted customs and practices tell you about the culture?
- **Understood (and accepted) terms of reference and rules for behaviour**. This isn't necessarily about what's *in* your Standard Operating Procedures, but more about *how* they are approached and utilised. It's about setting examples of good behaviour, and tolerance (or, hopefully, lack thereof) of poor behaviour. Do people always put on their safety gear before they enter the production area? Or do you see people fiddling with their earplugs once they're 20 or 30 metres across the shop floor? Such terms of reference permit regulation of the culture, and serve as tools for continuous improvement through application and adherence to rules and process.

- **Explicit and implicitly applied symbols which foster a sense of membership and aid identification of the group**. In simple terms, this includes the logos, slogans, colours, catchphrases and imagery utilised to project and communicate your 'brand'. Membership symbolism can operate at many levels within the organisation, from the macro level symbols that identify the organisation and what it stands for, to the micro level symbols that reflect local groups, departments or teams. (For example, do the Quality Assurance Lab staff all wear white coats and safety glasses? Do First Aid assistance providers wear a distinguishing badge? Do Fire Wardens or Marshals wear reflective or high visibility clothing? Are production teams recognised by their shift pattern ('the Morning Team'), by simple identifiers ('the A Team') or other features ('the Blue Team')? Is there a corporate uniform or dress code? These explicit symbols may all encourage a feeling of shared identity, but often more subtle symbols may be at play. One of the world's biggest science companies uses the colour red in its corporate logo and marketing materials. Although no corporate uniform exists, whilst walking around their global headquarters recently, I observed how the majority of people wore some type of red-coloured item as part of their own dress choice. A tomato-coloured tie here, a pair of scarlet shoes there; tiny details such as elegant claret-coloured cufflinks, sparkling rubies on fingers, earlobes and dangling from chains around necks, and, of course a proliferation of crimson-tipped fingernails. With a smile, I wondered whether this was corporate cultural influence at its most subtle, or whether these people simply recognised the wellbeing benefits of the colour (red is the colour most often associated with happiness and wellbeing[1]), or whether it was simply the law of averages at play and statistically more people simply prefer that colour.[2]
- **Shared values**. These are the moral standards or principles that serve to guide and provide direction to us as individuals, groups or organisations. When shared, values translate into the prevailing attitudes and beliefs within the organisation. Whilst easy to write down and publicise on a policy or poster, values are much harder to see in practice, and it may not be until things start to get a little rocky before the test of adherence to values becomes real. Authentic values are the 'corporate culture compass'. In order to be truly shared, leaders need to be in routine communication with the workforce to show how their activities and roles connect them to the organisational values. So, for example, if one of the values is 'we always work in safety' during team talks, shop-floor walk-arounds or in project meetings, leaders could ask "how does your role ensure we all work in safety today?"

These hallmarks, spanning attitude, language, behaviour, beliefs, rules and visual identifiers are not presented as definitions of culture, but more as 'defining elements' – the DNA of culture itself.

By considering culture in this way, as 'DNA', the guiding force of how we do what we do, we can begin to consider how to assess, shape and build our own organisational cultures in a logical and systematic manner.

REFLECTION POINT

Thinking about your organisation, can you identify the 'hallmarks' of your corporate culture? Take a moment to consider each and note them below.

- What are the buzzwords, phrases or common language used?
- Which customs and practices have become the norm?
- What behavioural rules are set in place?
- What symbols are used to help people identify with the organisation?
- What are the shared beliefs and values in your organisation? How do these influence and shape attitudes at work?
- What other observations can you make about your organisation's culture right now?

The psychology of culture

We all need emotional stability, whether at home, at play or in the workplace. This stability is key to an organisation's culture. Connecting each of the elements of the 'cultural web', just like the spider's silk, are the 'psychological contracts' employees have with the organisation.

At the heart of the relationship between every employee and the organisation are the perceptions and beliefs – the 'psychological contract' – that act as the foundation for the work agreement. The psychological contract is principally concerned with the relationship between the employer and the employee, specifically the mutual expectations between the two parties. Like the silk, these contracts are hard to see, but they do matter, however they appear oft-forgotten and worse, can be frequently ignored in the day-to-day cut and thrust of doing business.

Edgar Schein originally defined the psychological contract in the 1960s as an implication that: "there is an unwritten set of expectations operating at all times between every member of an organisation and the various managers and others in that organisation . . ." (Schein, 1985).

Over the years, the definition has broadened, principally to underscore the significant two-way subjectivity around these 'unwritten expectations': "the employment relationship consists of a unique combination of beliefs held by an individual and his employer about what they expect of one another . . ." (Armstrong, 2006).

It's crucial to note this subjectivity as the psychological contract essentially hinges on interpretation. Interpretation from two perspectives – the employer, and the employee.

In the same way, we can expect an element of reciprocity, of balancing 'give and take'. This reciprocity is just like the consent of an employer to pay a certain amount in salary to an employee for carrying out specific tasks or delivering certain results. Except, unlike the formal contract of employment – a visible, written manifest, typically stating obligations, rights and conditions such as working safely, remuneration and localised rules – the psychological contract (PC) is, by stark contrast, not so clearly defined. Instead, it's unwritten, invisible and based not on specific requirements or particular rules, but reflects the actual relationships built and sustained between the organisation and the individual worker. As each employee is a unique individual in their own right, this means that unlike formal employment contracts (EC), there is no potential for duplication or generalisation. Each psychological contract is therefore also unique, and what's more, completely fluid, dynamic and subject to the changing nature of the employment relationship, and the emotions that develop within it.

Expectations around corporate social responsibility have grown exponentially in recent years, and societal sentiment has begun to affect and influence the relationship that the employee has with their employer, in the sense that the employee generates a sense of 'community pride' for the work she undertakes.

Accordingly, when the organisation performs an act that benefits or supports the society in which it operates, this builds a positive relationship influence into the employee's side of the psychological contract. Conversely, if the organisation is found to have damaged society in some way, for example through an industrial accident, or environmental issue, the contract is negatively influenced.

This relationship influence (RI) can be seen clearly from the employer-society tie. Using the same examples, an industrial accident, or environmental issue, consider how each affect the organisation from a social perspective. Imagine the impact on share price following a chemical spill to the local river, or media attention gained after an employee is killed at work. Whilst we can understand the impact at the level of reputation management, the subtle cultural webs that form psychological contracts – whether between employee, employee and employer or those which tie influence between the parties and society – are less tangible, but no less important to consider.

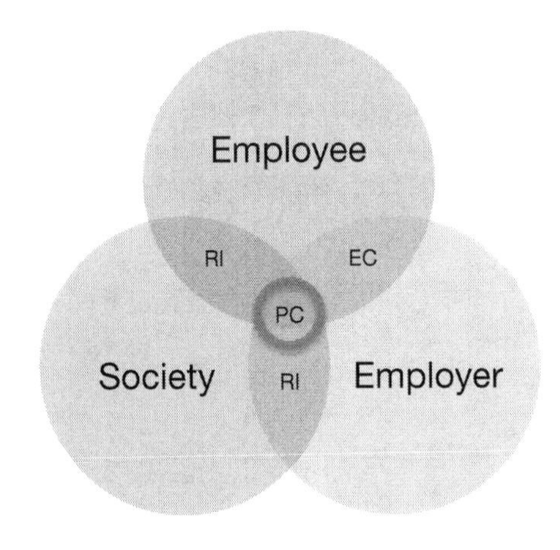

RI – relationship influence
EC – employment contract
PC – psychological contract

Figure 7.1 Contract–influence interplay

So, what are the beliefs that underpin the psychological contract? Trust, faith, transparency, promise, commitment, loyalty, communication, following through – each of these, and more, serve to influence employees' daily contributions to the workplace. A breach of trust, continued rudeness in communication, a boss not supporting a team member, an undelivered promise, a lack of recognition – such subtleties deeply affect the way in which people carry out their roles and the equitable reciprocity or 'give and take' of doing business.

Psychological contracts are the finest, most invisible threads in the spider's web of corporate culture. They not only help tie and bond individuals and groups together, but also shape the culture itself. Myriad studies underline that engaged workers are happier workers, and that happier workers are more productive and gain greater job satisfaction and sense of worth, so it makes good sense to think about how to build engagement within the workforce. How we *feel* about our work is perhaps the most powerful dynamic governing how we *do* our work. Strengthening the psychological contract is key.

The perils of a weak corporate culture

Even organisations with great products and services in high-growth industries fail to meet their expected success when leadership has allowed a poor culture to cultivate and take hold.

There can be any number of revealing signs that the presiding culture is unhealthy – ranging from displays of one-upmanship; a high-blame ethos; stereotyping; backstabbing; a lack of accountability; low morale; absenteeism and poor communication.

This sort of situation is, sadly, relatively common and tends to occur when organisations become preoccupied with growth, market share and products and, thus, lose sight of the importance of people, culture and sustainability. Though such a focus may produce short-term results, it tends to see organisations struggle in the longer term.

When the German company Daimler – makers of Mercedes-Benz – merged with US automotive company Chrysler in the late 1990s, it was initially referred to by the two companys' public relations teams as a 'merger of equals'. However, since the companies had markedly different cultures, it wasn't long before cracks began to show.

Differences included their level of formality, issues relating to pay, holidays and expenses and leadership style. Aware of the obvious tensions, the leadership decided they must try to infuse a single ethos across the company. However, what was finally agreed, was much closer to Daimler's culture – with the result that worker satisfaction levels at Chrysler plummeted.

A contemporary industry joke at the time was:

> "How do you pronounce DaimlerChrysler?"
> "It's easy: 'Daimler' – the 'Chrysler' is silent."

For the next few years the two previously successful companies limped on together, dogged by losses and forced to make cutbacks. Finally, less than a decade after the merger, the two companies parted ways.

Culture exists in every organisation, either by accident or by design. In more progressive organisations – especially those looking for long-term success and sustainability – leaders go to great lengths to nurture and positively direct it.

Beyond workplace accidents and ill-health, organisations regularly face a number of problems when culture is ignored or allowed to fester, two of the more significant are:

- **Increased costs:** when organisations are inequitable and fraught with in-fighting, there are always repercussions, such as delays, additional meetings, overturned decisions, greater levels of bureaucracy and duplicated work. All of which result in a decrease in productivity – which, in turn, increases costs.
- **Staffing problems:** a negative workplace culture is always liable to lead to problems with staff – such as poor communication, lack of compliance and decreased morale. Unhappiness, stress and a lack of teamwork will inevitably lead to higher instances of absenteeism and a higher staff turnover.

Rather than writing culture off as simply a 'soft' part of corporate life, good business leaders – like Gerstner at IBM – recognise that it is the element an organisation swims in. If not fostered properly, it can become turbulent, unmanageable and, ultimately, destructive – as DaimlerChrysler found to its peril.

Encouraging a strong corporate culture

So, what can we do to create a positive culture at work? There are four key pieces of the jigsaw to focus attention on: Values, Accountability, Respect and Feedback. Let's look at each of these now.

Values

Business leaders need to identify and continually communicate and promote the core values of the organisation. These should not be aspirational or 'nice-to-haves' – an organisation's core values are their fundamental guiding principles, their 'corporate compass', the 'Governing GPS'. They represent what the organisation stands for, providing direction and a moral guidance for the people that work within it.

Core values should also provide the basis for consistent and – hopefully – correct decision making. An organisation's values should speak volumes about who they are, both individually and collectively.

In defining a set of corporate values, rather than selecting terms that sound 'professional' or might win the popular vote, it's essential to select genuine ideals

– concepts that describe what the organisation's leaders believe to be essential facets of operational excellence.

I have visited too many organisations that have confusingly selected 'fun' as a corporate value, only to discover that they are actually run in a very regimented and compliance-heavy manner. When there is such a disconnect between your espoused 'core values' and your organisation's ethos, there seems little point in having them.

Accountability

Regardless of industry, geography or market share, I suggest that accountability should feature as a core value for every organisation.

Accountability creates trust – an essential part of any relationship – and being accountable demonstrates an organisation's commitment to be responsible for its own actions. Not only that, but it also reduces the time workers spend on distracting activities and engaging in unproductive behaviour.

Simply adding the word accountability to your corporate values statement is not sufficient, though. Accountability is action, rather than aspiration. We'll explore this further when we discuss aspects of behaviour and leadership, later in this book.

Respect

According to a recent survey of 20,000 workers by the Harvard Business Review, nothing has a more positive impact on worker behaviour than being treated with respect by management.

Forget compliance, training, mentoring, money or benefits – all of these things pale into insignificance when compared to being respected by a leader. Workers respond well to being treated decently – and, evidently, it puts them in a positive and productive state of mind.

On the other hand, of course, a lack of respect fosters resentment. Let's be clear, here: a lack of respect doesn't always involve harsh words or chiding, it can manifest itself in all manner of subtle, even subliminal, ways. It might even be picked up on when it's not actually there – so management must remain vigilant and be aware of just how much impact their words and body language can have in the workplace – and try to keep in mind the need to remain approachable, personable and engaged with workers.

However, this is not always as easy as it might seem, after all, how does one go about 'showing respect' for someone else? It's not the same thing as simply singing their praises – and whilst many leaders do heap on praise, if it's given too readily it quickly loses value. On the other hand, if it seems bland, perfunctory or jars with their tone of voice or body language, discerning workers will quickly perceive that it's disingenuous.

To truly cultivate a workplace culture that embraces respect is never going to be a 'quick win' – it takes time, effort and consistency of purpose. It also requires organisational leaders to treat people at all levels within the organisation with courtesy, politeness and kindness – regardless of what sort of day they're having.

Treating workers with respect doesn't mean that leaders are obliged to never disagree or have a cross-word with any of their staff, it just means that need to keep their arguments focused, measured and tempered with civility. Debate is healthy and positive – whereas, petty recrimination and refusal to engage, is not.

Feedback

In organisations with strong corporate cultures, feedback will always be expressed (and received) in a positive manner. That is not to say that it will be given with all manner of inappropriate or excessive blandishments – it will be accurate, factual and relevant – and delivered in such a way that reinforces what the worker did correctly first, before then identifying what improvements should be made. Repeated criticism rarely fails to inspire anything but infuriation and belligerence.

Feedback works best when it relates to a specific goal – and is delivered in a timely fashion. Often the best ways to get feedback across is for it to be done in situ. For example, an individual working with heavy machinery, will obviously need considerable 'on the job' training in order to work efficiently and safely. This kind of training needs to be carefully planned to ensure that it is thorough and covers everything that it needs to.

When feedback is part of a performance management programme, individual and team performance naturally improves, enhancing the performance of the organisation. With effective feedback processes, workers achieve higher levels of focus – and find it significantly easier to reach goals successfully.

THE NAKED TRUTH: CORPORATE CULTURE

Though it is often written-off as a 'soft' element of corporate life, the inescapable truth of the matter is that culture matters.

In cultivating a robust and positive culture, organisational leaders need to be able to show that they have the confidence to invest in the commitment and success of their workforce. This leads to increased morale, higher engagement and stronger performances.

Research by the Corporate Leadership Council (2004) has determined that workers are 87 per cent more likely to remain in a role they are happy and engaged with. This is important because staff turnover is costly, time-consuming and can be hugely demotivating for workers.

Additionally, the most recent annual *Best 100 Companies* report by *The Sunday Times* newspaper (February 2017) showed a trend in new job-seekers actively seeking out and prioritising positions based on factors surrounding culture – such as organisational ethos and values. So, evidently, culture has become a powerful differentiator in an overcrowded marketplace – and a way of attracting talent.

There can also be a significant financial benefit for organisations in successfully forging a positive culture. Surveying corporation Gallup recently performed an extensive study which polled more than 1.4 million workers around the globe – and subsequently discovered that organisations with positive cultures and high levels of worker engagement were 22 per cent more profitable than those that did not.

GET NAKED: CORPORATE CULTURE

1. Think about the corporate culture in your organisation. How would you characterise it? How do you suppose it came into being? Was it more by chance or by deliberate design?
2. Corporate culture begins at the top. What techniques do your organisation's leaders employ to help cultivate a healthy and positive work environment? How would you rate employee engagement and communication?
3. Consider the four elements of positive culture: *Values*, *Accountability*, *Respect* and *Feedback*. Identify how the organisation approached each of these factors – how did they do? Where were the good practices? And where was there room for improvement?
4. Think about a time when something went wrong or not according to plan. How was this managed? Based on that incident, would you say that the organisation displayed more of a 'learning' culture or a 'blame' culture?
5. Have you ever been part of an organisation where the top leader's personality shone through with strength and clarity? What were the particular personality traits of this leader? What nuances or mannerisms did they have? Can you recall moments where this same personality seemed to appear within teams, departments or other individuals? Why do you think that was?

Notes

1. Eva Heller (2000) *Psychologie de la couleur – effets et symboliques*, p.46.
2. Around 45 per cent of the global population prefer the colour blue, and whether for cars, living room walls or the latest fashion, red rarely makes it into the top five, so, it seems in this case, the colour could well be an implicitly implied symbol of membership for this particular organisation. And, yes, I was also wearing a red tie that day too!

8 Change

Driving change in an established culture can be a tough task for any organisation.

Currently around 88 per cent of all corporate change projects fail (MacPherson, 2017). It doesn't matter whether it's a plan to enhance corporate culture, deliver a new organizational strategy, IT project, merger or acquisition: only one in every eight succeed.

With democracy in crisis, globalisation causing business disruption around the globe, and nanotechnology and Artificial Intelligence increasingly threatening the way we work leaders struggle to implement the changes they know are necessary. Typically, they'll articulate their new vision, adapt policies and make the necessary new appointments – but still nothing substantive occurs.

So, in this chapter, we'll consider why organisational culture is so resistant to change, and explore the best methods to achieve it – to make required changes actually take effect. In the fast-paced modern business environment, being able to positively effect organisation-wide change can spell the difference between an organisation's success and failure.

The global competitive landscape has changed significantly in recent years, forcing organisations of all shapes and sizes to regularly respond to and engage in change activities.

Accordingly, the need for clear and competent change leadership is urgent. Whether Chairman, CEO, Vice President, director, senior manager or supervisor, the role of the leader is to bind groups together and enable them to perform. Success is rarely dependent upon the latest technology or on a specific business model. Instead, it typically comes down to the leadership's ability to visualise the organisation's future, forge a path to it and navigate their way through any roadblocks along the way.

Leadership can play many roles within an organisation, and in times of organisational change, these varying roles can be employed as different means to different ends – but regardless of the type of collaboration environment, it is always leaders that hold a key role as the agents of change. In these current challenging and competitive economic organisational landscapes, the need for strong leadership is increasingly evident.

Leadership must clearly, confidently and credibly articulate a vision, and direct, align and inspire action across the organisation. Without this, the organisation is reduced to a list of confusing, mismatched and time-consuming projects moving in disparate directions.

Creating a new order

Niccolo Machiavelli held a strong view on leadership in times of organisational change. Five hundred years ago, the Italian philosopher, writer, humourist and all-round Renaissance man, famously declared: "There is nothing more difficult to take in hand, more perilous to conduct, or more uncertain in its success, than to take the lead in the introduction of a new order of things" (*The Prince*, 1532).

Today, it remains one of the major challenges in the corporate world. How does one go about introducing 'a new order' of things, if history, traditions and mindsets are so well defined?

Though it is rarely an easy task, it is possible to transform an organisation's established culture – but it takes planning and perseverance.

If we define culture as 'the way the things are done around here' then it's easy to see that – however it manifests itself – culture is always going to be something that its stubbornly self-preserving and resistant to change. It is essentially what is already there, ingrained.

Culture is dependent on several interconnected factors that may have evolved over several years – such things as the organisation's mission, goals and values, as well as the dominant attitudes and behaviours.

Whilst it might be fairly easy for an organisation's leadership to change an organisation's policies, procedures and even realign its goals, objectives and ambitions, none of these things will bring about a cultural change.

To instigate any real change, a clear plan is required. But before that, it's important to strip things back and start at first principles. Leaders need to begin by asking some basic questions. Here's a starter for ten:

- What is the current situation and what needs to change?
- What are the pros and cons?
- What will be the new capabilities delivered by the change?
- What performance measures and targets are to be utilised?
- What are the expected improvements?
- What will success look like?

With answers to this first step gained and understood, the leadership team need to unite behind whatever the new direction is – and then create a robust strategy for action.

Whilst it's down to leaders to articulate and promote an organisation's vision and values, when it comes to changing its culture many of them simply assume this will happen automatically once they've articulated the new direction. In reality, this is far from the case. In fact, it typically needs to work the other way around – the behaviour needs to be the first element that is addressed. Only when inroads are made there, will organisations stand a chance of making a culture change.

As we've read in previous chapters, essentially, an organisation's culture is the collective behaviour of its workers. This is why influencing it may often take years, and why changing it can be problematic.

As well as being creatures of habit, human beings are social animals. In groups they will naturally polarise and create social norms – often at a subliminal level and unknown to themselves.

However, by approaching it correctly, a change in culture can be achieved – even in extremely well-established organisations.

Looking for culture

To attain an insight into an organisation's culture, it's important to attempt to view it from the outside.

Next time you visit any organisation – large or small – take a step back, and try to observe it as an impartial witness. How do workers interact? Are there conflicts? If so, how are they resolved?

How does the leadership team interact with the rest of the organisation? Are they visible, or do they cut themselves off in offices?

Look at the behaviour of the workers: do they seem engaged, animated, relaxed and friendly? Or do they appear to be isolated, despondent, stressed or withdrawn?

Look at the physical lay-out of the organisation: is it open-plan? Or heavily demarcated?

Observe the work and common areas: are they inviting or sterile? What things make up the environment? Are there family photos? Plants? Knick knacks? Cuddly toys?

It's also important to look for what *isn't* there.

If workers are ignoring obvious safety violations, for example – are there reasons? If no one leaves their workstations during the lunch period, what might that be indicative of?

Leaders need to lead

Changing an organisation's culture begins (and ends) with the commitment of its leaders. If any member of the leadership team is unclear on what needs to be achieved, or is wavering in their support, then the change is doomed to fail from the off.

At the same time, if leaders are unable to successfully articulate to workers the reasons behind the change, its relevance and why it represents a positive move forward, it is likely to experience serious problems.

Workers need to 'buy into' the change. If they don't understand the rationale, they are likely to become resistant – and what will start as confusion, will quite quickly transform into anger, unrest and opposition.

Leadership teams should establish a high sense of urgency in management and workers and provide clarity to develop coalitions. Plus, in the modern business world, the fluid nature of competition, and increased rate of change, underlines the need for continuous, rather than episodic leadership.

In my book *From Accidents to Zero* I talk of the four keys of evolution. To make an effective change, it's helpful for leaders to keep in mind these steps when attempting to drive a change in behaviour:

- Encourage
- Engage
- Enable
- Empower.

Encourage

From infancy, human beings learn through encouragement.

Think about how you learned to ride a bike as a kid – if it was anything like the way I did, your parents probably encouraged you, explained that it would be fun (and a useful skill), bought you a bike and went out with you holding the bike upright as you frantically peddled around your garden or local park – until, of course, you felt confident to ride it yourself.

Encouragement is an innate driver of behaviour, and, consequently, in organisations without a culture of encouragement, workers are much more likely to underperform.

It is also an extremely efficient way of focusing attention on tasks and driving workers towards goals. All learning typically begins with encouragement – as it's proven to be an extremely effective motivator.

Encouraging workers can take myriad forms. From something as simple as creating a new process to make desirable behaviours easier to perform, to demonstrating recognition of a valuable contribution.

Engage

All too often, organisations – even ostensibly progressive, safety aware ones – invest large amounts of time, money and resource into developing complex safety improvement programmes, campaigns and tools – only to be surprised to find out that their seemingly robust, shiny, new programmes don't work.

Usually, this happens because they have neglected to consult their own workers – the same people that are in a unique position when it comes to correctly identifying issues, solutions and approaches. What gives them such valuable insight? Well, the fact that they are there doing the work every day.

Even an exceptional safety expert viewing the organisation is going to be limited by time and exposure to its processes – the workers, that do the work on a daily basis, will naturally see the fuller picture.

Engagement with workers is not only vital in establishing a clear idea of the key issues within an organisation – but, also, in fostering participation in any solutions.

Before engaging workers, it's always a good idea to consider the possible outcomes and think about what responses might signify a positive outcome – for example, discovering why one area of the organisation consistently eclipses the others in terms of reported accidents and near misses – and then carefully tailoring questions in order to provoke the best possible responses.

For more useful answers, it's a good idea to script some 'open' questions. Closed questions, like: "Do you feel as though safety is important in your department?" will most likely elicit a simple "yes" or "no" answer. Whereas, questions like: "What do you consider to the biggest risk to your personal safety at work?" are likely be answered in a far more considered and revealing way.

Questions that could be asked to discover more about an organisation's safety culture might include:

- "What do you consider to be the organisation's attitude towards safety?"
- "Why do you think safety is important?"
- "When does your team discuss safety issues?"
- "Where could safety be improved?"
- "Who would you report a safety concern to?"
- "How could you go about improving safety here?"

What other questions could you add to this list?

An old technique taught to journalism students when learning how to interview people is to keep in mind '*The Six Ws*' – 'What,' 'Why,' 'When,' 'Where,' 'Who' and 'hoW'.

Reporters are taught to use these when conducting interviews as it helps them attain a fuller picture and, in journalistic circles, they are considered basic prerequisites in information gathering.

Obviously, they shouldn't be used robotically – or even necessarily in order. When engaging with workers, it's important to try and be naturalistic – if you're too formulaic, people will pick up on this, and may assume you're not really interested and simply 'going through the motions'.

Questions don't all need to be positive either. Often useful responses come from asking about negative aspects of the workplace, as these could be perceived to be more direct and getting to the heart of the matter. So vary your focus, perhaps by trying questions like "What aspect of your working life do you consider most likely to cause you to feel least safe?" or "What do you consider to be the more dangerous part of your role here?"

Enable

Having encouraged and engaged workers, it is important to then 'enable them' by providing them with the necessary skills, knowledge and tools to work safely.

Enabling is much more about instilling workers with the requisite competence and confidence to perform well – rather than, as it is sometimes thought, simply delegating work.

For example, it might be about ensuring that the Personal Protective Equipment the workers use on a daily basis is still fit-for-purpose and comfortable, or that employees feel confident performing tasks inherited from other workers following the streamlining of the organisation, or organising 'away-days' that promote better teamwork across departments.

Successful training is also an effective enabler – and this not simply concerned with briefing workers on what needs to happen or how it should be done. It should be about creating changes in behaviour.

Training is not the same as imparting a basic education either. Scott Geller, the prominent American safety culture expert, often highlights the difference quite effectively.

He often starts his seminars by asking: "How many of you have teenage kids? How would you feel if their school called you up and announced that they weren't going to do sex education any more – they were going to do sex training instead!"

Geller also has an interesting view of how organisations should go about retraining following periods of great upheaval and change – and that is to train the willing first.

Following any major change, organisations should effectively ignore those workers that are resisting the change and focus on those that have responded positively. Geller believes that the best way to get good results is to first work with the willing – then 'set them loose', as he puts it, on the rest of the workforce.

A key aspect of enabling people to support a change is for the change leaders to consistently and continually explain why the operations, task or goals have changed – and ensure that all workers are fully cognisant to the reasons why those changes are taking place.

Following the 'shock of the new' that comes with any organisational change, there will always be a period of worker apprehension, and leaders need to expect this and recognise the reasons for it. In *The Anxiety of Learning* (2002), organisational culture expert Edgar Schien asserts that people are always anxious in times of change and that this hinges on two conditions – the anxiety to survive the change, and the anxiety of learning something new in order to survive the change. In order to manage worker anxieties, leaders must develop strategies that communicate and reinforce why the change is positive. Trusted leadership teams will find this process easier.

Empower

There is a far stronger correlation between worker involvement in safety and incidence rates than between compliance and incidence rates. It is not hard to see why. After all, it's the difference between knowing not to perform a certain action for a specific reason – and just being told you can't do it.

Compliance and control are not sufficient behavioural motivators. Organisations need to move beyond using them as 'the goal' – and find ways to encourage communication and facilitate empowerment within the workplace.

Systems and structures within the organisation also need to be aligned to support and facilitate the desired change – and workers need to be made to feel as comfortable and positive about the change as possible. This is the only way they remain accepting of the change.

Leaders should also provide further support to further expedite this – providing such things as practice environments, coaching sessions and ongoing (positive) feedback. If the dynamics of the learning process are understood, the process of creating a new culture doesn't have to be a long, protracted affair.

Leaders need to provide workers with empathy and clarity of vision, then deliver a structure that helps them make sense of change (Brown and Eisenhardt, 1997).

Communicating change

Any organisation dependent on a human workforce, needs to have effective communication systems in place. Having the capability to send information, exchange ideas and issue (and receive) commands efficiently is crucial to the smooth running of any organisation.

When workers feel more engaged, it boosts morale, creating greater levels of workplace satisfaction – which, in turn, leads to greater levels of productivity.

As is elaborated on in the chapter on Behaviour, and often stated in books and journals about safety, most of the research pivots on the fact that 'human factors' are the primary cause of around 90 per cent of workplace accidents. If we consider some of the more high-profile workplace disasters of recent history – Chernobyl, Bhopal, Challenger, Piper Alpha and recent events with BP at Texas City and Deepwater Horizon – it should be noted that, with all of them, human error was considered to be a primary causal factor.

However, even with excellent lines of communication, organisations often still run into trouble with their corporate messaging – especially when dealing with topics as subjective and open-to-interpretation as safety. For example, how one worker might construe words like 'hazard,' 'danger,' or, indeed, 'safety', might be completely at odds with how another does.

When communicating on matters of corporate significance, safety leaders will often – quite naturally – revert to their own personal vocabulary and select words they consider to be appropriate replacements for the corporate message. However, there is often a very real danger that these words might be misconstrued by the audience.

Safety communications are often fraught with complications. After all, what does 'potentially unsafe' mean? Does it mean 'there's an outside chance that someone could sustain an injury'? Or does it mean: 'if it's not promptly dealt with, it has the capacity to cause mass fatalities'?

Further difficulties arise simply because many messages about elements in the modern work environment routinely get punctuated with jargon, overly-technical language and confusing abbreviations.

Consequently, it's always essential when making any kind of corporate communication to consider demographics, psychographics, cultures and dominant workplace 'attitudes,' in order to tailor the content as much as possible – ensuring that the language, style and delivery methods are appropriate to the context.

Whether in times of organisational upheaval or stability, creating and delivering clear, concise and consistent safety communications that will resonate with target audiences is essential. When organisations achieve this, they enhance the credibility of workplace safety whilst, at the same time, promoting the human skills behind safety leadership.

When trying to effect a change of culture, marketing campaigns need to be wide-reaching and ongoing. They also need to persistently promote desirable behaviours – and to do so in a way that is authoritative without being preachy and, if possible, supported by impartial material that backs up the positive reasons for change.

However, not all communication will – or should – be official, and leaders need to remain vigilant and 'on-message' at all times – even with regards to non-verbal communication.

The everyday behaviour of leaders needs to be consistent – and be the model of the behaviours they want emulated by the workers. In short, leaders need to be positive role models.

THE NAKED TRUTH: CHANGE

The constantly evolving landscape of modern workplaces is fraught with competition as barriers to cross-border trade are removed and companies find themselves operating within global marketplaces and in ways they never before considered. Change has become the new constant and corporate survival now depends on being able to out-manoeuvre, out-behave and out-perform peers. In short, it depends on the ability to change.

The majority of change attempts fail. Despite what may appear to be the creation of a clear vision, robust plans and solid milestones and metrics to measure and assure progress, only one in every eight corporate change programs succeed. It seems that Machiavelli was right when he remarked that there was nothing more difficult than to take the lead in times of change.

So, what makes that one in eight a success? The answer is clear. It's *people*. The anxiety that change brings sets people off-balance as they wonder whether they will personally survive the impending change – and whether they actually have the skills and knowledge needed to survive and then thrive into the future.

Leaders have a crucial role as the agents of change, and on their shoulders rests the success – or failure – of the intended change effort. In modern times, perhaps the focus needs to be more on evolution, rather than revolution – carefully building on what's gone before and served well, whilst concurrently setting in play the necessary elements to drive further improvement – instead of grandiose talk on the urgency and necessity of change. A series of solid, steady actions to systematically engage people on the change, encourage their participation, enable their contribution and empower them to succeed all serve well to reduce apprehension and anxiety, and build trust, belief and commitment in the new order of things.

GET NAKED: CHANGE

1 How does your organisation introduce, plan and respond to change? Does change appear without warning, or is time taken to fully encourage, engage, enable and empower people through the change journey?
2 Has your organisation ever experienced a comprehensive overhaul (such as downsizing, merger, acquisition or radical change to either market geography or products and services)? If so, how successfully was the change executed? Can you remember how workers responded to the change announcement? Were they positive or negative? Were they shocked or indifferent? What were the causal factors that created this response?
3 How open are your leadership team about major change? Is it something that they engage with workers about? Or something that is discussed privately behind the closed doors of an office or boardroom? How does your organisation encourage workers to contribute to change?

4. Journalists use the 'Six Ws' as a tool to ensure they capture everything they need to know about a story, but this tool can also help us on a daily basis in our dialogues on safety at work. Create a set of six questions using the tool that you can use during your interactions with employees today.
5. Identify a recent change in your organisation. Reflect on the communications that heralded the change. How would you rate the language used – was it clear, compelling and easy to understand? In what way did the communications impact the level of natural anxiety towards the change?
6. With regard to safety, what change would you like to see now in your organisation? What factors do you need to consider in order for your change to be embraced and sustained?

9 Safety culture

An organisation's safety culture is the product of the prevalent values, attitudes, perceptions and shared behaviours of its workers, and their collective influence on how safety is approached, managed and sustained in the workplace.

A positive safety culture is characterised by the endemic perception of the value of safety practices, improved workplace health and safety and enhanced organisational performance.

In this chapter, we'll explore safety culture, its definition and history, how it can be measured and assessed, why leadership commitment is so important to its ongoing development and how a weak safety culture can often lead to disaster.

Back in the 1950s, American anthropologists, Alfred Louis Kroeber and Clyde Kluckhohn, famously reviewed ideas and definitions of 'culture' – and compiled a list of 164 different definitions.

Six decades on, a conclusive definition remains elusive – and with countless definitions to choose from, opinion remains largely divided.

Ultimately, when discussing culture in the workplace, it relates to the collective grouping of an organisation to a code of conduct and mode of thinking in order to reach a set of prescribed objectives. What is often simplified in the workplace environment as: 'the way we do things around here'.

It's what American psychologist Arthur Reber described as "The system of information that codes the manner in which the people in an organised group interact with their social and physical environment" where the "frame of reference is the sets of rules, regulations, and methods of interaction within the group" (Reber et al., 2009).

Culture is the way that individuals act, prescribed by the structures and processes within the workplace, and directed by the values they hold dear.

In *Corporate Culture* (2012), psychologist Edgar Schein – who has devoted a considerable part of his career studying workplace culture – defines it as follows

> Culture is a pattern of shared tacit assumptions that was learned by a group as it solved its problems of external adaptation and internal integration that has worked well enough to be considered valid and, therefore, to be taught to new members as the correct way to perceive, think, feel in relation to those problems.

The term 'organisational culture' was popularised by business managers Tom Peters and Robert Waterman following the publication of their bestselling book *In Search of Excellence* (1982). In it, they argue that organisational success can be enhanced by culture – as long as it's decisive, empowering and people-oriented.

Peters and Waterman suggest that the culture of an organisation is largely invisible to individuals inside it, and that the values and beliefs are picked up in an unconscious or subliminal manner. Often workers are not even aware of the culture's existence – and, in order to get an accurate picture of it, it often requires observation from outsiders.

It is important to operational success that the culture fits with the demands of an organisation's objectives. Clearly, when the values of the prevailing culture are aligned it will hugely benefit the organisation. For example, a design company is likely to thrive if the embedded culture is one that promotes innovation, flexibility and worker empowerment – as this will naturally support performance. Whereas, within a culture of rigid conformity and respect for rules, it would more likely result in performance difficulties.

Dutch social psychologist Geert Hofstede states that culture is a set of unwritten rules of behaviour that set out what a particular group expects its members to do and what to believe. As a result, culture is typically a more powerful way of controlling and managing behaviour than an organisation's rules and regulations. For example, if an organisation experiences an unprecedented technical issue, it is likely that there are simply no rules to follow – so, obviously, a culture that demands strict adherence to procedure would struggle. Whereas, a culture that encourages creative or analytical thinking would naturally be better positioned to resolve such an issue.

In *Culture's Consequences: International Differences in Work-Related Values* (1980), Hofstede describes culture as the "collective programming of the mind within a group" – and suggests that an insight into an organisation's culture is fundamentally important to its "success".

If values and behaviours of workers are aligned with the objectives of the organisation, it will naturally lead to advantage. Where there is misalignment, it is more likely to result in performance problems and organisational failure.

Safety culture

Like culture, 'safety culture' has proven a difficult thing to define. Which means, of course, there are numerous definitions to draw from.

However, there are two that are used prominently. The following was produced for the International Atomic Energy Authority in 1991 by the Health and Safety Commission (HSC) in Great Britain: "That assembly of characteristics and attitudes in organisations and individuals which establishes that, as an overriding priority, nuclear plant safety issues receive the attention warranted by their significance." And one produced by the Advisory Committee on the Safety of Nuclear Installations (1993) in the wake of the Chernobyl disaster states "The safety culture of an organisation is the product of individual and group values, attitudes, competencies and patterns of behaviour that determine the commitment to, and the style and proficiency of, an organisation's health and safety programmes."

Whilst most contemporary organisations will happily acknowledge that safety culture is an important and valuable concept, without any real clarity or consensus, it remains a subject that often gives rise to conflict.

Psychologist James Reason suggests that, during workplace discussion about safety, few things are more reliably raised than 'safety culture'. Yet, despite the concept's apparent acceptance and popularity, it remains little understood.

The concept of 'safety culture' is still fairly new. It was only in the aftermath of the explosion at the Chernobyl nuclear facility in 1986 that 'poor safety culture' suddenly gained international attention – when it was identified by the International Atomic Energy Agency as the primary cause of the accident.

The Chernobyl disaster sent shockwaves across the world – and it continues to reverberate today. Exposure to acute radiation claimed the lives of 49 workers as a direct consequence of the accident. However, a more recent investigation by the United Nations has suggested that the overall number of premature deaths linked with the Chernobyl accident is likely to be more than 4,000.

Before Chernobyl, organisations tended to approach safety as a 'command and control' activity – usually compliance-based and focussed on strict adherence to rules and procedures – but the scale of the disaster clearly demonstrated that contemporary safety management techniques needed review.

However, the concept of safety culture has gained further acceptance in recent years, following the investigations of many other prominent accident investigations, including those looking into the causes of the Piper Alpha oilrig disaster, the Ladbroke Grove rail crash, the explosion of NASA's Challenger space shuttle, and the Macondo Deepwater Horizon oil spill – the investigations of which all identified poor safety culture as a key causal factor.

Britain's workplace health and safety regulator, the Health & Safety Executive (HSE), suggests that often the issues surrounding safety culture are due to misunderstandings by organisational leaders – with many assuming that if they have a weak safety culture it is due to the poor attitudes and behaviours of workers.

In a review of the literature the HSE concluded that:

> Many companies talk about 'safety culture' when referring to the inclination of their employees to comply with rules or act safety or unsafely. However, we find that the culture and style of management is more significant, for example, (. . . there is regularly . . .) a natural, unconscious bias for production over safety.
>
> (Gadd, 2002)

In *Developing an Effective Safety Culture* (2001) James Roughton and James Mercurio assert that leadership is the most important part in developing a robust safety culture; stating that "If management demonstrates a commitment and provides the motivating force, and the needed resources, to manage safety, an effective system and culture can be developed and will be sustained."

The point is further enforced in the conclusion of *Measuring Safety Climate: Identifying the Common Features* (Flin et al., 2000), psychologist Rhona Flin's meta-analysis of 17 studies on culture across a variety of sectors and countries. The article concludes by determining that organisational leadership was the primary source of safety culture.

Author of *Safety Culture: Theory, Method and Improvement* (2017), Stian Antonsen, concurs with the direction set by Roughton and Mercurio, and Flin's meta–analysis, and suggests that leaders can guide positive change on two levels:

- Influencing conventions for safe behaviour, interaction and communication.
- Setting values, norms and underlying beliefs and convictions through which workers deal with risks.

Where safety cultures are weak, organisational leaders need to become positive safety role models – because, at the end of the day, how safety is prioritised within an organisation is the responsibility of the management hierarchy.

Features of safety culture

Professor Dominic Cooper, one of the world's leading thinkers in behavioural safety, believes that safety culture can be considered the product of the following three interlinked features:

- Psychological aspects: individual and group attitudes, perceptions and values;
- Behavioural aspects: safety-related actions and behaviours;
- Situational aspects: policies, procedures, organisational structures and management systems.

Cooper describes the psychological facets as 'how workers feel' (what they value, what they think about their role, morale, etc.), the behavioural as 'what workers do' (their everyday functions, their interactions, the things they do to actively improve safety, etc.) and the situational aspects as 'what the organisation has' (safety management systems, policies and procedures, guidelines, safety personnel, etc.).

In order to optimise the safety in an organisation, Cooper asserts that it is first necessary to improve the situational environment – which will, in turn, positively impact behaviour. Over time, as workers adjust their behaviour, they will automatically alter their values to fit with the behaviour they are engaging in. As James Reason once notably put it "organisations, like organisms, adapt" (1998).

In *Drift into Failure* (2011), psychologist Professor Sidney Dekker takes issue with organisations that claim to have 'done' safety, because they have successfully brought down the numbers of accidents. He states that, in the workplace, the natural condition of people and systems means that it is in a constant state of 'dynamic flux' – and, accordingly, the number of unforeseen events or consequences is inestimable. However, rather than recognising this, as statistics improve, many safety practitioners become comfortable – and slowly drift away from excellent performance and into failure.

Dekker maintains that robust safety culture demands drive and ongoing advancement. It requires the systematic application of measures and behaviours, several ongoing methodologies and people-management techniques, that provide safety as a natural consequence.

Let's take a look now at three popular models of safety culture that may provide a framework for your organisation's cultural journey.

Schein's cultural model

Arguably, the most important – certainly the most cited – theoretical model of safety culture is Edgar Schein's three-layered cultural model. Schein states that there are essentially three component elements that shape and influence safety culture:

- Organisational Artifacts;
- Espoused Values;
- Underlying Assumptions.

According to Schein, Organisational Artifacts are such things that are readily observable in the workplace. This will include such things as architectural features or furniture style, branding, corporate symbols and graphics. It will also be evident in common communication styles, dress codes, rituals, ceremonies and how events are carried out. Typically, artifacts are tangible and recognised. However, the 'meaning' of organisational artifacts can often be difficult to decipher – though their impact might be significant.

Though Espoused Values are not necessarily directly observable, they can often be inferred from observing how workers behave – and, crucially, how they choose to represent the organisation (to themselves and to each other). Espoused values may lie within the organisation's stated beliefs and principles, or be expressed in official philosophies, mission statements and procedures.

Underlying Assumptions are the source of the organisational values and artifacts and form the essence of culture. They are present in workers' everyday behaviours, and are deeply-entrenched and typically carried out unconsciously. Since these assumptions are ingrained in any organisational dynamic, they are often extremely difficult to observe or identify. They are the workers' collectively understood, taken-for-granted beliefs, perceptions and thoughts.

The three levels of Schein's cultural model are often presented as being like the layers of an onion. Organisational Artifacts are depicted as the outer layer, as these are comparatively easy to observe and influence. Espoused Values, which are more difficult to perceive or control, are shown as an inner level. Finally, deeply embedded in the core of the onion, are the Underlying Assumptions – which are even more elusive and harder to observe.

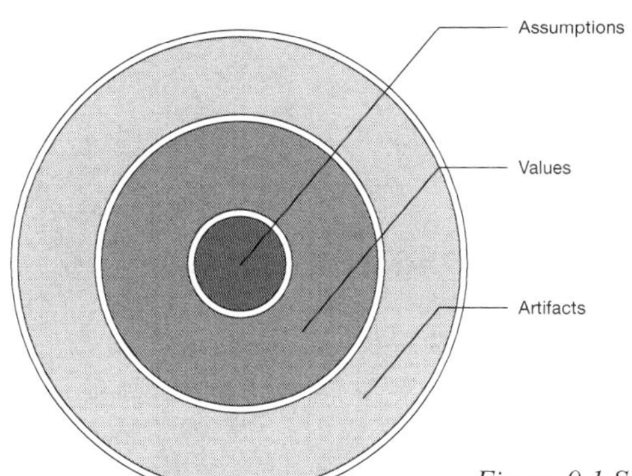

Figure 9.1 Schein's three-layered model of culture

The DuPont Bradley Curve

A popular model designed to help organisations understand and benchmark the journey towards a strong safety performance is the DuPont Bradley Curve.

Created by Vernon Bradley, whilst working for chemicals giant DuPont, the model proposes that, when it comes to safety, there are four different stages that organisations will move between during their lifetime. These are described as:

- Reactive
- Dependent
- Independent
- Interdependent.

The Reactive stage describes a situation where workers in an organisation consider that accidents will inevitably happen – and there is little that can be done to change the fact. In this stage, avoiding accidents is viewed as something that happens through luck, rather than design.

The Dependant stage is characterised by compliance. In this stage, organisations are more reactive than proactive – and use rules and regulations to enforce objectives. During this control stage, worker behaviour is dictated by the wish to avoid punitive measures, and the journey towards total safety – or 'zero accidents' – is considered a vision.

At the Independent stage, there is more understanding of safety processes and greater levels of worker commitment to improving safety within the organisation. There will be freedom to share ideas and an emphasis on self-improvement. Attaining 'zero accidents' is now seen as a difficult – but achievable – goal.

At the Interdependent stage, there is substantial cooperation throughout the organisation. The workers display high levels of pride in their work. Safety is a shared value – and workers are engaged, and will look out for one another. Attaining 'zero accidents' is now looked upon as an expectation.

Since the introduction of the DuPont Bradley Curve in 1995, it has been used extensively by safety professionals as a way to assess the safety culture of organisations – and to chart improvements.

Since many organisations rely on compliance-based safety systems to achieve safe behaviour at work, they often remain entrenched in the traditional Dependant stage for a considerable period. However, sooner or later, the law of diminishing returns will force them to update policies and procedures and proactively endeavour to create safety improvements.

Typically, this will see them either adding further levels of compliance (which can only really be a short-term solution, since it is likely to bring about the same situation further down the line) or make the far more radical choice of attempting to inspire workers and change mindsets. The latter choice bringing them into the Independent and Interdependent stages.

The Parker and Hudson Model

One of the fundamental errors that many organisations make is attempting to create a strong safety culture by simply focusing on improving systems.

As we've seen, safety culture is not a static concept. It has many working parts, each of which are continually – if, often, subtly – shifting.

These include the organisation's systems, procedures, attitudes and perceptions – and, essentially, it is the resulting interaction of these different elements that creates the organisation's safety culture.

Having state-of-the-art systems will naturally be a bonus for any organisation – but it doesn't mean that people will actively engage in safer practices. For this to occur, an organisation needs to develop a 'mindset' of values that are not only espoused, but are lived.

Developed by psychologists Diana Parker and Patrick Hudson, the Parker and Hudson Model is an assessment tool often used by safety professionals to evaluate the safety culture within organisations. The model works on the basis that organisations can be split into five distinct types:

- Pathological
- Reactive
- Calculative
- Proactive
- Generative.

Pathological – these are organisations in which safety is viewed as an expensive inconvenience. There is very little attention given to safety issues, and accidents are viewed simply as a cost of doing business.

Reactive – these are organisations that only respond to safety issues once an incident of poor safety performance has occurred. Whilst much may be done in the wake of an accident progress is slow, and occurs only in fits and starts.

Calculative – this denotes organisations that manage safety through systems and predominantly use lagging indicators to assess the safety of worker performance. (In this instance, safety is viewed of as a set of standards to be complied with.)

Proactive – these are organisations that look towards the future of safety success, actively involving workers in the management of safety at work. Focusing on the inputs to creating safety in the workplace, progress is measured through a balanced scorecard approach, using a blend of leading and lagging indicators.

Generative – these organisations are characterised by the possession of what is sometimes referred to as a 'healthy paranoia' for safety, manifested as people looking out for each other and where speaking up is embraced and encouraged. Safety is held as a value by all workers and everyone in the organisational hierarchy feels that they have a role to play in keeping workers safe and healthy at work.

The dynamics of safety culture

Whilst the Parker–Hudson model sits apart from the Bradley Curve through its empirical and academic pedigree, there is a similar risk of the model being perceived as a linear progression. These models should not be seen as the straight-line progression towards world-class safety. Rather, they should be considered as being more like stock markets, where the value of the investment – in this case, the safety of workers – can go down as well as up. Safety culture in itself is not a static concept. It is complex and dependent on multiple variables – such disparate factors as an individual organisation's systems, policies and procedures, taken in conjunction with its human workforce's behaviours, attitudes and perceptions. It is the interaction of these different factors that forms the basis of safety culture within an organisation.

Focusing purely on systems will not drive safety improvement. Whilst it is true that good safety systems should be viewed as a prerequisite it does not automatically follow that with better systems workers will engage in safer practices. Indeed, as we'll see, James Reason believes there is a distinct correlation between the implementation of strong safety systems and poor safety performance in workers.

To create a truly robust safety culture, organisations need to proactively set about making safety a value for all workers – and, to attain this, it requires significant commitment and a drive towards continuous safety improvement.

Safety climate

It is not unusual to hear the terms 'safety culture' and 'safety climate' used practically interchangeably – however, the two terms do actually refer to different things.

In their book *Safety Culture: Philosopher's Stone or Man of Straw* (1998), psychologists Sue Cox and Rhona Flin discuss both subjects – and describe safety climate as "a manifestation of safety culture in the behaviour and expressed attitudes of employees".

This chimes with Edgar Schein's view that "climate proceeds culture" – and that, in fact, climate is simply culture in the making.

Typically, climate is seen as much more superficial than culture – it is more like a description of workers' attitudes to safety. For example, though subject to the same policies and procedures, workers within an organisation will often react very differently to them, depending on which part of the organisation they are employed in.

According to Dominic Cooper, reactions to rules are predisposed to fluctuate throughout organisations depending on 'local' habits or practices. The perceived levels of risk in workers' job roles, for example, will naturally influence attitudes towards compliance. So, safety climate is more prone to variation in different parts of an organisation – whereas, culture is more intense and overarching.

Culture concerns the underlying beliefs and convictions of the whole organisation, often considered to be its 'values'.

It all boils down to people

In 1988, just off the coast of Aberdeen, Scotland, the Piper Alpha platform exploded, killing 167 workers stationed either on the rig itself or in one of the safety standby vessels patrolling it.

The controversy surrounding the disaster was intensified when a Public Inquiry, headed by Lord Cullen, contended that the operator, Occidental Petroleum, had used inadequate maintenance and safety procedures. The report went on to produce more than 100 recommendations on how safety could have been improved at the site and how disaster might have been averted.

Finally, the investigation exposed the fact that the disaster was caused by a simple and very human cause – lack of communication.

On the day of the accident, there had been a worker shift change, during which a vital piece of information (that one of the main pipes had been sealed with a temporary cover – and, as a result, had no safety valve) had not been relayed to the incoming team. As a result, gas had leaked from the pipe and ignited – causing the explosion.

Regardless of their contrasting industries, sectors, operating styles and locations, most of the prominent process safety disasters witnessed over the course of the last few decades have shown a weak safety culture to be the major contributing factor.

Based on the findings of the investigations into a number of major international disasters, Dominic Cooper compiled a list of common factors relating to poor safety culture disasters in his 2013 book *The Strategic Safety Culture Roadmap*.

Features typical to these disasters included:

- *Profit before safety*. Where organisations value productivity above safety; often resulting in safety being viewed as an expense, rather than an investment.
- *Fear*. Commonplace in organisations where safety is based on compliance, and punitive measures are used to reinforce behaviour – which often leads to workers covering up or failing to report accidents/near misses in order to avoid recrimination.
- *Weak safety leadership*. Where risks are not recognised or properly followed up – and opportunities to control or halt them are subsequently lost.
- *Miscommunication*: Where the channels of communication have broken down – and, consequently, essential safety information has not been conveyed to decision-makers.
- *Competency failures*. Where expectations that workers with a direct influence over the organisation's safety practices were either experienced or knowledgeable proved to be false.
- *Lost learning*. Wherein lessons that should have been learned from previous incidents were either lost or ignored – and, therefore, critical safety information was either not remembered, shared or enforced.

According to Cooper, if an organisation displays any of the features listed above, it needs to be quickly addressed.

However, should it display three or more, it needs to be resolved as a matter of supreme urgency – as Cooper believes that the potential for serious incidents occurring increases exponentially with the number of poor safety culture features an organisation has.

Creating an ideal safety culture

In *Achieving a Safe Culture: Theory and Practice: Work and Stress* (1998), James Reason states that an ideal safety culture should be "the engine that drives the system towards the goal of sustaining the maximum resistance towards its operational hazards".

Citing indifference, loss of focus and complacency as common causes of major accidents, Reason suggests that there is also an additional danger that happens when organisations invest in complex, defence-in-depth style safety systems.

He argues that when there are no accidents or near-misses to steer workers, they quickly lose their 'intelligent and respectful wariness' about potential dangers. This often results in a worrying – and dangerous – trend in business leaders believing that their organisations are 'safe' simply because they have no data to suggest otherwise.

When organisations lose sight of potential hazards, Reason claims that they often become preoccupied in other areas – such as production or sales – and all aspects of workplace culture connected to safety are duly allowed to slide. Soon the organisation becomes vulnerable, simply because its leaders believe safety has already been accomplished – and have become blindsided to deficiencies in workplace conditions and working practices.

Far from removing issues around safety from organisations, Reason believes that complex safety systems, of the type often used on major hazard sites, regularly become the catalysts of poor safety culture. When it comes to safety, complacency is demonstrably a killer. Reason believes that an 'ideal safety culture' can only be achieved if workers keep an open-mind to potential threats, treat them with due reverence – and remain 'intelligently afraid'.

As Pablo Neruda memorably observed: "You can cut all the flowers, but you can't keep spring from coming."

Leading from the top

Establishing and sustaining a strong safety culture in an organisation is key to creating safety. The myriad influencing factors and maturity levels certainly present distinct challenges to the practitioner. So where to begin?

When it comes to safety, the behaviour of an organisation's leaders and senior management team is always extremely telling. This is because the strength – or weakness – of a safety culture is almost always mirrored by the level of leadership commitment.

Typically, organisations with poor safety cultures will be those that barely communicate about safety – they will be far more concerned with products or profits. As a result, safety will rarely – in any meaningful way, at least – find its way onto the corporate agenda. It will rarely be a part of any of the leaders' personal interactions either.

The behaviour of leaders and senior managers – and their level of engagement – is critical to how safety is perceived throughout an organisation, as it sets the tone on how the other levels of management should behave – which, in turn, filters down to the rest of the workers. Without setting a clear and positive 'tone from the top' on safety, it is unlikely that the safety culture of an organisation will mature.

A popular phrase for organisations keen to improve their performance is that 'safety is everyone's responsibility'. However, this is only true in organisations where safety is recognised as a core value. In organisations with poor safety cultures, you'll usually find innumerable workers (from the boardroom to the shop-floor) that will point out, in all seriousness, that safety isn't part of their job.

It is the role of committed safety leaders to consistently promote the importance of safety, and to inspire a positive attitude towards it, by setting the example workers are expected to follow – and by recognising good safety behaviours.

THE NAKED TRUTH: SAFETY CULTURE

If we view organisational safety in its historical context, it's clear that over the last few decades there has been a growing emphasis placed on the cultural and behavioural aspects of safety management.

Successive investigations into major disasters such as Chernobyl, Piper Alpha, Kings Cross, NASA's Challenger shuttle and Deepwater Horizon have revealed that despite complex safety systems in place, accidents occurred due to human causes.

However, none of these incidents were simply a matter of individual human error. Instead, they were caused by a lack of safety vision embedded within the attitudes, values and perceptions dominant throughout the entire organisation. To put it simply, they were all organisations with poor safety cultures.

To avoid such harmful cultures developing, organisational leaders need to stop talking about safety as a 'priority' – or even a 'top priority' – and, instead, need to uphold it as a 'core value'. Priorities change, whilst values hold fast.

Once leaders have established safety as a core value, they need to communicate this throughout the organisation, not through edicts and dispatches from ivory towers, but through personally becoming visible safety role-models by consistently engaging with workers about positive safety practices, and treating safety as one of the key factors that determine organisational success.

Overwhelmingly, research shows that organisations with positive safety cultures have lower accident rates and workers who perceive leaders to be committed to safety engage in more safety-related (and fewer risk-taking) behaviours.

GET NAKED: SAFETY CULTURE

1. As you read this chapter, how did the definitions of safety culture impact you? Was there one definition in particular that stood out for you? How does this definition compare to how your organisation views safety at work?
2. How do leaders in your organisation influence conventions for safety through their own behaviours, interactions and communications? Consider that their influence may be deliberate or unintentional, positive or negative. Are there certain leaders that have a particularly stronger influence than others?
3. Consider the psychological aspects of safety culture – what attitudes (of individuals and of groups or work teams), perceptions and values exist in your organisation when it comes to safety?
4. Which situational aspects (such as policies, procedures, structures and systems) in your organisation can you identify that support the growth of a positive culture? Which might restrict progress?
5. What are the espoused values of your organisation when it comes to safety? What does your organisation (and its senior leaders) believe? What is the predominant philosophy for safety? Check your corporate mission statements, policies and procedures for ideas and then consider whether the organisation is truly living these in practice. If not, why not?
6. Review Cooper's list of common factors in accident disasters that have safety culture at their root. Which of these might present in your organisation?
7. Look at the stages of the Bradley Curve or the Parker–Hudson model of safety culture. What is your initial feeling as to where your organisation sits on these models? What do you need to focus on now in order to mature to the next stage?

10 Behaviour

Since back in the 1950s when Heinrich concluded that up to 90 per cent of serious workplace injuries are caused by human behaviour, the figure has remained a constant in most safety books and journal articles. Yet, it's data that is perhaps less surprising when you consider that almost all accidents involve people.

Whilst robust Safety Management Systems and sound policies and procedures have prevented hundreds and thousands of workplace injuries over the years, many safety teams have found themselves on a performance plateau – and, as a result, have chosen to focus on what they perceive to be the panacea: influencing behaviour.

In this chapter, we'll explore some of the prevalent research into behaviour, from Skinner's operant conditioning studies in the 1930s, to the recent popularity of Nudge Theory, in order to provide an overview of how important it is to consider the role of human behaviour in workplace safety.

We'll discuss ideas to encourage positive action in the workplace and also seek to understand the limits of human behaviour.

When looking to attain excellence in organisational safety, it is vital to comprehend the nature of human behaviour and what drives human choices.

In high-performing safety cultures, there is always an appreciation that, whilst there are many influences over behaviour, none is as significant as the expectation of what the consequences of an individual worker's actions will be for them personally – and, by extension, for their colleagues and teams.

Behaviour is largely an ingrained function based on how individuals – either directly or subliminally – perceive consequences. When organisations appreciate this fact, it means that they can exert significant influence over the behaviour choices of their workers. In relation to safety, this is especially useful when it comes to risk-taking activities.

It's not unheard of for organisations to quickly embed habitual practices throughout the workplace that amount to 100 per cent safety critical behaviours, often manifested as 'Golden Rules of Safety'. But, before they can do this, it is necessary to fully comprehend the 'consequence drivers' that foreshadow individual and group choices.

Since most organisations approach production and quality entirely methodically in an effort to maximise performance, it is odd that this isn't exactly the approach used when it comes to safety performance management. In practice, many organisations abandon scientific rigour and simply rely on the training and experience of their safety personnel – which is, of course, valuable – but more often than not, unlikely to include the fundamental tenets of behavioural science. When statistics from the OSHA in the US and the HSE in the UK suggest that 90 per cent of workplace accidents are behaviour-based, it seems like blatant oversight to ignore the science behind the challenges we seek to resolve.

Consequence drives behaviour

Human behaviour is, as you already know, influenced by myriad factors. If we consider why a worker acts in a certain way, it is likely we will come to understand that their motivations will come not only from the tasks they have been set (and, presumably, accepted to perform), but also influences based on the culture and structure of the organisation, and, of course, by their own personality and physical condition.

Each of these factors has a part to play in achieving an understanding of the activities within the organisation – as they all have an influence on what behavioural science will tell you is the driver behind almost each and every choice: consequence.

Rarely do workers engage in unsafe behaviour just for the sake of it. Normally, there will be a system of formal and informal checks and balances that goes on inside their heads that results in them believing it's a good idea – i.e. that the outcome is worth the risk. The reasons for such behaviour are varied and might include one or more of the following.

- It may be something that would positively enhance the worker's own experience to some degree – such as saving a few moments of time by not putting on protective clothing to complete a minor task, or cutting corners on a tedious larger task in order to finish it more quickly.
- Alternatively, it may be that a worker has weighed up the consequences of their unsafe behaviour – and believe it is unlikely to get noticed, or, otherwise, if it is discovered, unlikely to result in any meaningful penalty. It might even be out of habit – perhaps the worker has accidently taken part in some risky behaviour and got away with it and then carried on, feeling that the safety guidelines are redundant.
- Finally, it might be simply that following safe procedure may actually conflict with what they believe the organisation wants. Perhaps they work in an area where high production rates are – or, at least, are perceived to be – the priority over safety? Is it impossible to reach quotas without engaging in unsafe behaviours? Is it further reinforced by line managers turning a blind-eye?

Focus on consequences, not antecedents

When safety practitioners mull over incident reports and ponder why certain events took place, often their focus is on *why* a worker behaved in a certain way. Whilst understanding why is important, there may be a risk of becoming stuck looking at the antecedents – or, rather, what occurred to 'set off' the behaviour. Examples of antecedents include speed limit signs or speed cameras on a highway; a sign on a garden gate warning of a dog; even dark clouds in the sky. All of these things tend to encourage a specific behavioural response. Whilst it is no doubt useful to consider antecedents, a focus here may cause us to miss vital aspects. Though it may seem counterintuitive, a deeper dive to examine the potential range of consequences, may help attain a clearer, more accurate picture of why a specific event occurred.

Take, for example, a worker that has sustained an injury to the hand due to not wearing the protective gloves that she was issued. If we examine the event, we can see that the injury occurred because she was working in a potentially dangerous environment and wasn't wearing the required protective clothing. In terms of antecedents there are signs on the factory walls advising that PPE must be worn. A mention of safety gloves is included in the Standard Operating Procedure for the task. How might one go about ensuring it doesn't happen again? Remind the worker to wear their gloves? It makes sense. However, thinking about it, that was already a rule. Presumably, the injured worker already knew that she was supposed to be wearing protective gloves. So, our next conclusion might be either that she was careless on the day, or she habitually flouts the rules. Either way, reinforcing the rule seems like a weak response – and one that is likely destined to fail again.

If, on the other hand, we actually try to get inside the head of the worker and think about why they might have gone about breaking the rules in the first place – i.e. reviewing the implied consequences of doing so – we would be more likely to get a clearer idea of why the behaviour (of not wearing the protective gloves) occurred in the first place.

Let's have a look at what the likely thought-process might be in this instance:

- Wear protective gloves – *consequence*: possible, but not definite, reduction of injury.
- Don't wear gloves – *consequence*: greater comfort, increased performance at tasks that require manual dexterity, quicker performance time.

An examination of workplace consequences can reveal that in many instances not operating in a prescribed way can be preferable to the worker, because, in many cases, the implied 'threat' involved – in this case, suffering a hand injury – seems vague and uncertain.

It's much like smoking, in this regard. After all, it's universally acknowledged that smoking is bad for our health. Yet, people still smoke. Why? Well, because the behaviour around smoking is not motivated by the implied consequence, simply because it lacks immediacy. Whilst individuals may understand that cigarettes are bad for their health, they are also aware the consequences are unlikely to take place instantly – they may believe that they have time to give up/get fit later.

Reinforcement

Reinforcing consequences are those that occur following any positive consequence to make the preliminary action/behaviour more likely to occur again in the future.

Whenever an individual uses a light switch, for example, there is a behaviour (pushing the button) and a consequence (the light blinking on). Assuming the individual pushing the light switch wanted the consequence that followed the action, this is likely to lead to repeat behaviour.

Every action has a consequence – a fact that is essentially the basis of all science and how the human brain marks the passage of time. It is also Newton's third law of physics. Understanding consequences is fundamental to our understanding of the world, so let's consider the 'carrot and stick' of reinforcement now.

Negative reinforcement

Much of human behaviour is based on avoiding things. Negative reinforcers are those consequences that individuals actively seek to avoid. It is a major factor in motivating human behaviour. If it wasn't, umbrella sales would slump.

Acting to avoid adverse conditions or punitive measures is an important part of human life and society. There would be no rule of law without it. But negative reinforcement – the 'stick' or threat – rarely inspires great or creative performance – and does not motivate to improve. When individuals are motivated by fear, they usually just toe the line – but, clearly, behaving in a certain way because they feel compelled to, will result in worse performance than if an individual behaves the way that they actually want to.

However, working under negative reinforcement conditions is the traditional and often accepted way, and, consequently, the system remains. In many organisations, spanning a vast array of sectors, work is performed adequately – simply in order to satisfy a compliance standard and avoid negative consequences.

Positive reinforcement

In the corporate world, it is less common to see individuals working under positive reinforcement conditions. Fortunately, though, the tide is slowly beginning to turn – as we discussed in the chapter on Engagement in this book – many leaders are now beginning to realise that a more positive approach is needed – especially if they want staff to go that extra mile.

Research shows that workers under positive reinforcement conditions:

- work because they want to;
- are more positive (working towards something good – as opposed to avoiding a negative);
- work more proactively and creatively.

Positive reinforcement promotes strong, durable behaviour change – and can take any number of forms such as acknowledgement, praise, money or anything at all that is provided after our behaviour.

Good leaders invest effort to identify their workers' positive reinforcers through understanding what motivates them, and by trying out potential reinforcers, for example, increasing positive feedback, providing 'stretch targets' or offering reward and recognition.

Behavioural-based Safety

Recent research by DowDuPont – formerly Dupont – the American chemical conglomerate, recently asked the question: *Is Behavioural-based Safety (BBS) enough?*

Whilst the theoretical principles of BBS are grounded in what have been shown to positively influence worker behaviour and increase safety within organisations, it does seem that a complete focus on how workers act in the workplace has limited effectiveness in the long term and it would seem that the initial reduction in safety incidents begins to plateau.

DowDupont consider there to be two reasons for this:

- **Habituation**: this is described as the tendency in all living organisms to stop responding to stimuli after prolonged exposure to it. This is, for example, why prolonged use of a recreational drug will naturally result in the loss of the initial 'buzz' – and why many addicts will start upping the dosage. In the workplace, when workers habituate to the environment, they often begin to quickly become immune to enforced regulation and safety communications.

- **Conflict**: because BBS is driven by external factors – such as fear of repercussions and consequences – it means that there is essentially a trade-off going off in the workers' heads. They will take part in safety processes, mainly at the behest of their leaders, supervisors or safety practitioners, but they are little more than an unwilling accomplice in the safety of the organisation.

DowDupont's research suggests that the most effective approach to workplace safety is to combine the principles of BBS with other psychological drivers. If there is any chance of a workplace becoming entirely safe, it would be necessary for workers to *choose* to be safe – and in order to this to take place it is likely to need 'an in-depth learning journey'. This would involve workers getting a better understanding of their individual safety – or 'independence' – and the safety of the workers around them – their 'interdependence'.

In the world of cognitive psychology, it is understood that much of what happens to shape human behaviour goes on a deeper, subliminal level. When working in an organisation, much of the behaviour of workers is dictated simply by operating within its social context – and much of that is governed by the leadership that oversees it.

Within any organisation, the prevalent values and beliefs make up the social norms which in turn then exert influence over the people working there. In organisations where safety is an important part of the culture, it becomes part of this social influence and quickly becomes an entrenched value, important at both the individual and group level.

A holistic approach

So, whilst BBS is a good method for improving workplace safety, when organisations use an approach that also utilises the social and cognitive elements this will create a robust, more sustainable, 'holistic' safety management process.

All organisations, regardless of industry, will face risks associated with the physical, social and leadership elements of workplace safety.

- **Physical** – the danger of injuries arising from work processes or from the working environment.
- **Social** – the risks of human behaviour and the tendency of workers in a group setting to shape to beliefs of the majority and adapt behaviour to the rules of the group.
- **Leadership** – the risks associated with a lack of strong, visible commitment to safety.

Many safety programmes focused on just one facet of safety, such as attaining zero accidents or enforcing heavy compliance-based operations, but when they focus on singular aspects of safety, there is generally little chance of sustainability.

Rather than providing workers with an endless stream of safety communications or demanding that they follow rules in some thoughtless, sheep-like way, safety programmes should try to factor-in situational elements and individual beliefs and attitudes to influence how workers behave in the workplace. This is what is sometimes referred to as 'situational awareness' and it is how strong safety programmes encourage workers to realise the importance of safe conduct – in real time and in the moment.

The goal of any truly progressive safety programme should be to align thinking with action and social context – for only when these factors are successfully integrated will the programme start to work, and the organisation be able to understand – and, crucially, influence – worker behaviour.

Learning theories

Behaviourists, like American psychologist B.F. Skinner, believe that complex behaviour is learned gradually through the modification of much simpler behaviours.

Imitation and reinforcement play important roles in these 'learning theories' – which propose that individuals learn by duplicating behaviours they observe in others. To reinforce a desirable behaviour, Skinner argues that it essential to reinforce it with a 'reward' of some kind. Positive reinforcement, as we have mentioned earlier. As each simple behaviour – for example, remembering to put on a hardhat in a construction site – is established through imitation and reinforcement, the complex behaviour develops (being mindful of safety).

In the early 1930s, Skinner developed a device known as his 'operant conditioning chamber' – better known today as 'The Skinner Box'. This was a glass case with a base capable of delivering small electric shocks. It also contained switches which could be pressed by subject animals, also capable of delivering small electric shocks. There was a light in the box, and a small chute which fed into the case from the outside. When the switch was pressed – *and* the light was on – this would deliver food.

Skinner's test animals – mainly rodents and birds – quickly learned to associate the pressing of the switch with either painful punishments (negative reinforcement in the form of electrical shocks) or pleasant outcomes (positive reinforcement, in this case, food), depending on whether the light was switched on or not as they pressed the lever.

Skinner's approach led to the ABC model of behaviour management – which contends that in order to influence behaviour, it is essential to first understand what motivates it. He also believed it was important to accept, from the off, that it is very hard to force people into doing things – especially if they don't want to do them.

As simple as ABC

The Activator–Behaviour–Consequence – or ABC – 'model of influence' is certainly one of the more popular theories relating to the psychology of safety and health culture.

Skinner believed that classical conditioning – like that proposed by Pavlov about his famous dog – was too simplistic to be a complete explanation of the complexities of human behaviour. He asserted that a lot of it was based on causal responses to previous actions and that human behaviour was based on consequences. Thus, if the consequences were good, the actions that led to it being repeated becomes more probable – whilst, conversely, if the consequences were bad, there is a high chance the action will not be repeated.

By way of simple example, if an individual picks up a hot coffee pot and pours themselves a nice cup of coffee, they are likely build on that pleasant experience and do it again in future. However, if they were to pick up a hot coffee pot and burn their hand, then they are probably less likely to repeat the action in the same way. Here we have examples of both positive and negative reinforcement.

The popularity of Skinner's ABC model is probably down to the theory's straightforwardness and that it is very good at explaining the 'lead-in' and 'follow-up' facets of behaviour.

The model involves the application of operant conditioning which is there to reinforce desirable behaviours. It proposes that people always need an activator – what we might otherwise refer to as an 'antecedent' or a 'trigger' that drives a particular type of behaviour. The activator shapes an individual's attitude and motivates them to undertake a specific course of action.

The modern world is filled with activators of all kinds – from the flashing lights at pedestrian crossings, to the male/female signs on toilet doors, to the close-circuit security cameras that festoon modern shopping malls. Each of these things prompt specific types of behaviours taken by an individual as a direct response to the activator. The consequence is whatever the result is of the specific action.

To influence behaviour then, individuals need to be provided with appropriate activators and consequences to encourage them to think for themselves whilst reacting in the desired fashion. Such as:

- Activator: Signage advises goggles must be worn on the factory floor
- Behaviour: The worker puts on protective goggles
- Consequence: The worker's eyes are protected when dust is ejected from a machine.

Why smokers smoke

When it comes to building a strong organisational culture, all three of the ABC factors are critical.

Activators without obvious consequences rarely drive behaviours. For example, in the UK, an old – but still active law – dictates that it is "illegal for the keeper of a place of public resort to permit drunkenness in the house". Anyone that has ever strayed into a British public house around closing time, will know how little this law is acted upon. Pub landlords, and their clientele, have clearly judged – quite correctly, it seems – that the risk of any specific consequence of breaking this law is very low – thus, the effectiveness of the activator in influencing their action is significantly reduced.

Now consider a behaviour without a specific activator. Imagine the heady days before 1964 when the first anti-drink-driving advert aired on British television with its – slightly sexist, albeit slightly telling – 'Don't ask a man to drink and drive' strapline. Before those flickering black and white TVs brought the message home, it was quite normal for people to have a drink and drive – and, indeed, the notion of 'one for the road' is still a fairly familiar one.

Worried by the growing spate of traffic collisions, the Ministry of Transport commissioned a 40–second animated film for television broadcast – three years before a legal drink-drive limit was even set. The advertisement's gentle reminder to not drink (because it might lose drivers' their driving licenses) is in sharp contrast the UK government's recent *Think!* campaigns – which tends to use shock tactics to show how drink-driving might, in fact, lose drivers their freedom – or, indeed, their lives!

However, that first advertising campaign was well-heeded – and in the succeeding 35 years the UK saw a six-fold reduction in drink-driving deaths. Evidently, that campaign – and the ones that were to follow – was instrumental in making drink-driving a social taboo.

In everyday life, we've largely grown to accept the myriad prompts and reminders that invade our lives. Whether it's a 'No Smoking' sign or a warning about speed cameras, activators can be a quick and powerful influence our behaviour in a positive way.

Yet, overexposure to activators can have a wearying effect. In historical terms, the internet is still a fairly young phenomenon – yet, the overzealousness of online marketers desperate for clicks has already resulted in something called 'banner blindness'. Web-savvy people have become so resistant to all those 'Click Here' links, that they're effectively immune to them. Indeed, if the 'heat maps' of web pages created by online consultancies are to be believed many users are so switched-off by these ads, that they literally can't see them anymore.

Other activators seem to have little effect for other reasons. For example, for around two decades now in the UK, it has been a legal requirement for cigarette packets to come with large and impressively scary-looking warnings of the danger of incurable cancers and impending death because of smoking. Yet, people still smoke.

Why? Well, because behaviour is not motivated by the *implied* consequences – these lack immediacy. Whilst individuals may understand that cigarettes are bad for their health, they are also aware the implied consequences are unlikely to take place instantly – they may believe that they have time to give up and get healthy later, for example. And enjoy the nicotine hit in the meantime.

In organisational safety cultures that are based heavily on compliance, the focus is usually on activators. This may prove useful, as workers can be given appropriate training and tools – and the consequences of non-compliance can be impressed upon them from day one. This gives workers reason to perform in a certain way and adopt specific behaviours. However, it is when workers become accustomed to their tasks and relax into their roles that problems start.

In much the same way that individuals will naturally feel more relaxed and less self-conscious when they are alone, workers will often relax into their tasks when they perform them under a perceived lack of pressure.

Social psychologists have long understood that when tasks are not immediately familiar or very complex, workers will typically work harder – and the effect is amplified still further when they believe that they are being observed.

In rule-based environments, the 'consequences' might take a number of forms. They can be:

- soon or delayed
- certain or uncertain
- positive or negative.

However, even with strongly-articulated consequences – like those adverts on the cigarette packs – the threat of negative consequence is unlikely to completely stop unsafe behaviours. Anything that is slow, inconvenient or boring – such as working on a production line – will often lead to workers cutting corners with a similar belief to the smokers, that something bad will never happen to them.

People power or cold showers?

Behaviour is also strongly influenced through reinforcement brought about by social pressure. When workers become associated with a particular team or group, they quickly fall under the influence of its membership – which requires conformity to the group's behavioural and attitudinal norms. Whatever the group adopts as 'normal' becomes promptly implanted in the common understanding hivemind of the group. This can obviously be either a good or bad influence.

An example of this having a negative influence involves a recent investigation into safety on commercial airlines which highlighted that within the close-working crews onboard, social pressure was extremely powerful.

The investigation discovered that on many aircraft the crews had cut the pre-flight safety demonstration down significantly in order to save time – and, in some more extreme cases, dispensed with it entirely. This had quickly become the social norm onboard and, accordingly, when members of the team did perform the demonstration in full – as company rules dictated – the group applied 'social sanctions' to the individuals that had deviated from the group's 'norms' and followed standard safety procedure. There's a classic experiment that illustrates the power of deviation from established social norms very well, involving a group of monkeys and a bunch of bananas, let's take a diversion to consider this here.

REFLECTION POINT: THE WAY IT'S ALWAYS BEEN AROUND HERE

A cage contains five monkeys, at the top of the cage a banana is hung from a string, and a set of steps is positioned beneath the banana. Fairly quickly, the monkeys will notice the banana and one will climb up the steps to reach the treat. As soon as he touches the steps, all the other monkeys are sprayed with cold water.

A short while later, another monkey dares to make the same attempt to get the banana. As soon as he reaches the steps, all the other monkeys in the cage are sprayed with cold water. As you can imagine, they are not happy chimps.

At this point, one monkey is removed from the cage and replaced with a new monkey. It doesn't take long before he notices the banana and makes a start for the steps. To his surprise though, before he reaches the steps, he is attacked by his peers and prevented from reaching the banana. After another attempt is met with the same response from the other monkeys, he understands that if he tries to reach the banana he'll be assaulted by the others, so he joins them on the other side of the cage, away from the treat.

Now another of these original monkeys is removed from the cage and replaced with a new monkey. As he enters, he notices the banana and makes for the steps. Just like before, he's set upon by the other monkeys – including the recent newcomer. The experiment continues, with a third original monkey being replaced with a new one. He gets to the steps and is set upon by the group – however, this time, two of the four of the monkeys taking part in the punishment have no idea why they weren't previously permitted to climb the steps and take the banana – and also no idea why they are beating this new monkey.

The trial progresses, replacing the fourth and fifth original monkeys with new ones. Now all of the original monkeys that were sprayed with cold water have gone, but despite this, none of the current five ever approach the steps – even though they want the banana. Why not?

Because as far as the monkeys are all aware, this is the way it's always been around here.

The tale of the monkeys and the banana explains well the power of social norms, how certain beliefs or attitudes can be cultivated within group settings, and just how culture – *the way we do things around here* – is formed.

A less obvious peer pressure, the perception of 'cool behaviour' – either exhibited by a dominant person within the team or in a certain clique – can often have a remarkably powerful effect on the social fabric of a larger group and exert significant influence. Indeed, a startling proportion of car accidents involving young drivers in the UK cite social or peer pressure as a key causal factor.

However, peer pressure can also be a force for good in terms of safety. The Health and Safety Laboratory's paper *Young People's Attitudes to Health and Safety at Work* contains several interviews with school-leavers discussing safety practices in their first jobs.

Many of the interviewees are recorded grumbling about being made to follow boring and longwinded safety protocols, however, one young worker seems far more occupied with the hardhat that they have been instructed to wear at work – and complains that every time that he attempts to remove it: "everyone's shouting at you 'get it back on, get it back on!' It slows your work down half the time but what can you do about it?"

As the stories of the monkeys and of the young lad on the construction site reveal, we need to address prevailing attitudes and beliefs in order to encourage positive social norms. To influence behaviours to create safety at work, it is important to engage with workers to find helpful activators to promote positive behaviours. And then follow this by providing positive consequences that reinforce that behaviour.

Just one more thing . . .

The COM-B model of behavioural theory – nicknamed the '*Colombo* theory' after the late Peter Falk's shabby-raincoat-wearing TV detective – proposes that there are three things needed to drive a particular behaviour:

- Capability
- Opportunity
- Motivation.

They're exactly the same things that the eponymous sleuth would look for when searching for a killer in the hit US TV show.

Often safety campaigns and interventions in organisations fail simply because they don't create sufficient motivation for workers. Just as the 'SMOKING KILLS' signs emblazoned across cigarette packets fail to make tobacco smokers change their ways in any significant numbers, simply because the claims are too abstract and warnings of 'potential peril' are broadly unmotivating. Weirdly, despite dealing with a potentially lethal prospect, these signs are demonstrably less motivating than a 'Wet Paint' sign on a park bench – simply because the latter deals with the here and now, rather than an obscure possibility somewhere in the future.

However, where there is successful motivation, workers still require the opportunity to carry out behaviour. There is no point in advising a worker to put on protective goggles, if there are none. Or explaining to someone the health benefits associated with mountain climbing if they have no access to a mountain.

Then, finally, there's capability. It's difficult to change behaviour if workers are unable to carry out what's being asked of them. So before embarking on a quest to change behaviour, look first at the capabilities of those you seek to change.

Managing negativity

Negative attitudes and behaviour at work can lower team morale and damage the performance of groups. By far the worst kind of behaviour within an organisation is negative behaviour that is both chronic and endemic. If a team is exposed to this kind of activity, it will often derail operations and can ruin any long-term strategy.

At the end of the day, workers are people, and everyone feels frustrated at times, or has the occasional stressful off-day. Yet, often the problem runs deeper, and this can be far more detrimental to the cohesion of teams.

When negative influence takes hold, it brings with it feelings of learned helplessness – and suddenly everything seems problematic and nothing constructive is happening. The positive attitudes promptly disappear, and it feels as though any feelings of joy that might have previously been in the team have been comprehensively sucked out.

In his book *Managing Workplace Negativity*, Gary S. Topchik cites research from the U.S. Bureau of Labor Statistics which estimates that organisations lose around $3 billion (USD) a year simply due to negativity amongst workers.

Topchik believes that in order to truly manage negative behaviour, it's important to deal with the root causes rather than the symptoms. I completely agree. All too often, on impulse, managers focus on attempting to address the visible behaviour – when really they need to look deeper and explore why a worker is acting in such a negative manner.

Whilst it might be tempting to rebuke the worker for undesirable attitude or behaviour, this is only likely to exacerbate matters. Rarely, in these instances, do two negatives make a positive.

There are several prevalent reasons for negative attitudes and behaviours in the workplace, that seem crop up persistently, quite regardless of industry:

- **Lost confidence** – often there has been a fairly imperceptible decline in a worker's personal confidence caused by being in the same role for a long time and they feel disregarded and 'in a rut';
- **Feeling unappreciated** – similar to the lack of confidence and often happens when a worker has been overlooked for a promotion – or some other change in status – and believe their contribution is undervalued;
- **Feeling lost** – this is when workers feel removed from the day-to-day processes of the organisation and feel displaced and powerless;
- **Stirring up change** – in some instances, workers will complain about working conditions simply because they genuinely believe that it is the best way to effect change in the organisation;

- **Job anxiety** – especially in the current economic climate, workers will be deeply anxious about job security and this will manifest itself in negative behaviour, often without them necessarily being aware of it.

There are, of course, other pertinent reasons why managers might want to carefully proceed when it comes to seemingly negative workers. There may be any number of unknown psychological or situational factors – often not related to work – that the manager might not be aware of and that the worker has not felt it comfortable to disclose. For example, there could be underlying mental health issues – such as depression or stress; or perhaps difficult personal circumstances, such as the worker facing a divorce or grieving.

It is essential for the manager to help the worker to understand that their negative behaviour is having a detrimental effect on the rest of the team. Once they've understood this, it's vital that the worker does not feel ostracised from the rest of the team, and feels able to re-engage effectively.

Most crucially, managers need to remember that these things happen – and learn from the experience. Over time, they will quickly – and intuitively – learn to realise the early warning signs and be able to put in place their own preventative measures to minimise the risk of negativity reoccurring.

Give it a nudge

In recent years, the British media have been reporting – often with some quite demonstrable confusion – about a government-adopted scheme that purports to use behavioural science to positively influence behaviour.

Called 'nudge theory' the idea is laid out in-depth by economist Richard H. Thaler and law professor Cass Sunstein in their best-selling book *Nudge: Improving Decisions about Health, Wealth, and Happiness* and is described as using research in psychology and 'behavioural economics' to influence behaviour without coercion.

Advocates of nudge theory – which uses a mixture of psychological persuaders and situational factors – claim that it is more effective than direct action, legislation or legal enforcement. Often used within safety situations, 'nudges' tend to be targeted towards specific groups of individuals and used to promote desired behaviour by means of mental influencer – using much subtler 'under the radar' techniques.

For example, traditional techniques to get motorway drivers think about their speed, would be by adding flashing signs advising them to 'slow down,' or by reminding them at intervals about the speed limit. However, many motorways are using a nudge technique, and are using radar speed guns that let passing drivers know what speed they are travelling at, without further comment. Research shows that this sort of psychological 'nudge' – in this case, to keep within the speed limit – is much more effective.

Arguably, the most often cited example of a nudge is the now infamous 'Dutch fly'. Without wishing to get into too much detail here – and relying on your ability to think creatively – the gents' toilets at Amsterdam's Schiphol Airport were subject to some fairly horrendous abuse from patrons, and so the management team wondered how they might go about motivating a change in their behaviour. After consulting a behavioural psychologist, they replaced the urinals with ones featuring a picture of a housefly just above the drain. The results were clear – as most urinal users suddenly became eager to 'dislodge' the fly in the bowl – and, in direct consequence, their aim improved and Schiphol's toilet cleaning costs massively diminished.

Thaler and Sunstein describe nudges as:

> Any aspect of the choice architecture that alters people's behaviour in a predictable way without forbidding any options or significantly changing their economic incentives. To count as a mere nudge, the intervention must be easy and cheap to avoid. Nudges are not mandates. Putting fruit at eye level counts as a nudge. Banning junk food does not.

Since publishing their book, nudge theory has been supported by a number of world governments, including the Obama and Cameron administrations in the US and UK. In 2010, the latter even set up the 'Behavioural Insights Team' – better known as the cabinet's 'Nudge Unit'.

Despite its relatively short existence, the Nudge Unit has had a number of high-profile successes. For example, they switched how court fines were demanded – from impersonal letters insisting on the immediate settlement of court costs, to sending out personalised demands via text messages to the defaulters' mobile phones. This increased response rates from 5 per cent to 40 per cent practically overnight – and saved the UK government around £30m in bailiff fees.

There is clear scope for elements of nudge theory to be incorporated into workplace safety, and many areas where it might, quite quickly, have a positive impact. Such as:

- **Positioning** – optimising the organisational layout of workplaces, and items within them, in such a way that promotes worker engagement – and, thus, encouraging the creation of safety. For example, such as positioning the Personal Protective Equipment supplies at point of use.
- **Communication** – adding greater levels of personalisation to safety documentation to speak more directly to workers and promote engagement and emotional investment. (e.g. 'John, when did you last have your eyes checked? Get a free appointment today!').
- **Social reinforcers** – creating visual prompts, such as signage, posters and stickers, to remind workers about the organisation's safety culture and remind workers of the predominant social norms and practices: *'Are you wearing a hardhat? Everyone else is!'*

Though it is obviously gaining acceptance within many industries and sectors – and, in fairness, has been proven to have demonstrable positive impacts on behaviours – nudge theory is not without its criticism. Many have been quick to point out that that the beneficial effects of nudges have been mostly short-term – and, particularly in relation to safety – research needs to be conducted in order to attain a greater understanding of the psychological factors that can bring about more long-term behavioural changes.

Another potential negative characteristic of safety nudges could be they could distract the worker from their actual work, which in some workplaces could potentially be fatal. For example, in 2017, in response to a recent surge in deaths on those delightfully twisting scenic mountain roads in high in the Swiss mountains local governors erected a series of road signs that each state single words ('Attention!', 'Distracted?', 'Focus!') set in cartoon-style speech bubbles. The signs were designed to refocus drivers' attention on the route ahead. From a purely personal perspective I've certainly noticed my attention is drawn to the lurid signage, and the immediately attendant urge to quickly look back to the tarmac ahead. Judging by the frequency of on-road skid-marks and clipped verges at the same locations, I'm not so sure these nudges are having such a positive effect on everyone.

Conversely, in the way that the human mind quickly becomes desensitised to repeated stimuli, it seems likely that nudges will need to be regularly revisited and refreshed in order to stay useful.

Even despite the good results they can garner, many take exception to nudges at a more fundamental level, and have chosen to see them as kind of subtle mental manipulation which, especially when employed by governments, looks a lot like social engineering.

Human error – active and latent failures

It is essential that accident investigators and operational managers view human error in its wider context – and consider the various factors might have contributed to it occurring in the first place.

Only with a better understanding of the real root causes behind accidents, can organisations hope to improve safety processes and avoid similar future accidents. Concluding investigations simply with the assertion that the accident was caused by 'human error' and then setting preventive actions which include telling workers to be more careful or state 're-training required' are wholly insufficient.

In his book *Total Safety Culture*, Tim Marsh points out that blaming a single individual for an accident might be useful in absolving management of responsibility – but it implies that the accident was avoidable, and is likely to instil a culture of fear throughout the larger organisation. To achieve a more detailed picture of the various types of human error that might arise, it is a good idea to look at the individual and organisational factors that can influence them.

Britain's regulatory body for workplace safety, the Health & Safety Executive (HSE), categorises two distinct types of human error – 'active' and 'latent'.

Active failures

Active failures are the direct and immediate causal factors behind an accident. Typically, they are caused by front-line staff working with tools or operating machinery.

The HSE categorises them in the following way:

- **Slips and lapses**: errors that occur during the execution of the correct plan of action – usually as a result of interruption, carelessness or fatigue;
- **Errors of judgement**: occurring in circumstances during which a choice between two or more courses of action needs to be made – and an incorrect choice is made;
- **Violations**: deliberate errors caused by individuals deliberately breaching established protocols or procedures.

Repetitive tasks carried out without much consideration are vulnerable to slips and lapses if the worker is unengaged or their attention is diverted.

Research from *New Scientist* magazine shows that an average fit, healthy and well-rested human being will only concentrate on a specific task for about 45–55 minutes in every hour – and that's the *best-case* scenario. When tasks are repetitive workers are considerably less engaged in what they are doing and 'tune out'. At such times, workers on a production line or other repetitive task become more susceptible to slips and lapses of attention as they tend to quickly lose interest in the task at hand. To put this in context, let's take an organisation of 20,000 workers. Assuming every worker employed there was *fit*, *healthy* and *well-rested* – this would still represent 16,000 hours lost every single week.

So, if you factor in every stressed, tired, ill, hungover or, indeed, mentally unwell worker, it's clear to see that there is going to be a significant amount of time that every organisation loses to its staff simply tuning out from what they are supposed to be doing.

Errors of judgement

Errors of judgement generally occur when a worker is given unusual tasks to perform, or is, more generally, working under pressure. These might occur as either 'rules-based mistakes' or 'knowledge-based mistakes'.

Rule-based mistakes are made with confidence, when a different procedure is chosen due to a lack of proper situational assessment. Think of an English car driver pulling off a ferry in France and covering their error by complaining that French people drive on 'the wrong side of the road'.

Knowledge-based mistakes are the kinds of errors that occur due to the failure to correctly appreciate a situation or make an appropriate decision. In which, an individual has no rules or routines available to help them in an unusual situation: resorts to first principles and experience to solve problems. Sticking to the English driver example, in preference to asking a local French person for directions, they might instead rely on an out-of-date map to plan the unfamiliar route to their hotel.

Violations, though deliberate, are rarely wilfully obstructive. Typically, they occur from a worker's desire to perform work given constraints or in a pressured environment. For example, ignoring established safety protocols in order to finish a task more quickly.

There are three separate type of violations:

- Situational violations
- Optimising violations
- Individual violations.

Situational violations are usually characterised by situations in which workers perceive that they have no choice but to break the rules. For example, situations where despite their best intentions, workers have been unable to meet a specific quota for whatever reason – perhaps a fire alarm meant workers were away from their station for long periods during the day or specialist equipment broke down and was unable to be operated.

Optimising violations denote those times where rules are broken because a worker has deemed the outcome is really what is required. This is where non-compliance becomes general practice and the consensus seems to be that a specific rule no longer applies. This is regularly seen on motorways, where drivers regularly drive about the specified speed limit – and clearly feel as though they do so with apparent impunity due to a lack of meaningful enforcement.

In the workplace, these sorts of violations are often well-meaning, but misguided and often exacerbated by unwitting encouragement from management for 'getting the job done'.

Individual violations are usually undertaken intentionally for personal gain or satisfaction. One of the key drivers in life is control – and individuals often do what they can to gain control. This is natural, of course – and individuals often take a 'calculated risk' in order to achieve a desired outcome. For example, all those people that drink heavily at a late-night party, only to wake up early the following morning and drive to work, not quite sure if they are still 'within the limit'.

Latent failures

'Latent failures' are factors that influence human decisions or actions and contribute to active failures. They are often made by individuals whose tasks are removed from operational activities – such as designers, rule-makers and organisational leaders.

Examples of latent failures include:

- poor design of plant and equipment;
- ineffective training;
- inadequate supervision;

- ineffective communications;
- inadequate resources;
- uncertainties in roles and responsibilities;
- procurement decisions.

Latent failures are often seen as 'traps in the system' – the underlying and hidden aspects which viewed casually seem entirely harmless but can often lead to terrible events.

A latent failure might be, for example, having two different buttons on a control desk that look similar but have opposing operations. They are errors caused by system flaws, rather than character flaws.

Cognitive dissonance

Cognitive dissonance is a psychological term for a type of mental stress experienced by an individual holding two – or more – contradictory beliefs, ideas or values at any one time. Typically, it follows an individual performing an action that contradicts their personal beliefs or values, or when they are confronted with new information that contradicts their previously held beliefs.

In *A Theory of Cognitive Dissonance* (1957), American social psychologist Leon Festinger stated his belief that people naturally strive for psychological constancy in order to function in everyday life.

According to Festinger, human beings need internal consistency – and suffer from psychological stress if they don't have it. This then will usually move them to reduce this, so that 'normality' can resume.

Generally, how most people will approach avoiding cognitive dissonance is by altering existing cognitions, adding new ones or by actively seeking to reduce the perceived importance of whatever the dissonant factor is.

Festinger's research shows that most human beings are more likely to actively warp new ideas so that that they no longer contradict their pre-prescribed beliefs than go to the effort of developing them and adapting their attitudes and behaviours accordingly.

Coming from Scotland, whenever the Rugby World Cup tournament comes around, I will obviously back my national team – and, unlikely though it might seem, with contenders including South Africa and New Zealand on the pitch, I will genuinely *believe* in Scotland's chances of winning the prestigious trophy.

How am I able to do this? Well, mainly by mixing with other Scots with similar inclinations, reading jingoistic national newspaper articles, ignoring information about the sides from other countries, etc. Basically, because I *want to believe* in my national team, I find myself able to do so. I start to cut off avenues that might conflict with my swelling national pride and find myself suddenly able to believe all manner of spurious details about the players' and manager's skill levels as long as they are consistent with my own thinking.

Cognitive dissonance is also prevalent in the workplace and can be a significant cause of stress for many workers. For example, workers in safety roles are often coerced into supporting – or, at least, tolerating – policies that can be the source of considerable personal conflict. For example, when they have proposed potentially life-saving changes within an organisation and are informed that these not being actioned due to budgetary constraints, it can be the source of considerable anguish. It might be the case that the Safety Officer chose their career in safety due to a profound belief that they can save lives or make a positive impact in peoples' lives.

In many fields of work there are occupations where these sorts of conflicts arise because far from being just a job their work is really more of a 'calling'. These sorts of jobs are based on an overriding set of principles. The Safety Officer hearing that her recommendations are not being implemented due to a lack of financial resources is little different from the medical professional being informed that their patient will not be cared for due to a shortage of beds. When forced to face conflict – and the added pressure to tolerate it – any committed professional is liable to experience significant personal frustration, and this could become to the detriment of their working performance.

There are any number of situations that might cause cognitive dissonance in the workplace; often it occurs in workers who are asked to perform tasks outside their perceived skill set or level of training, as a consequence of poor leadership. It also regularly arises when workers are required to do something for the organisation that they genuinely believe to be wrong. For example, when a member of a Human Resources team is made to discharge a worker for some misdemeanour without any supporting evidence being submitted or under obviously mitigating circumstances, they are likely to experience significant cognitive dissonance. The conflict here is clear. The HR person has to, somehow, reconcile in their own mind, their individual principles with the expectations of a more senior decision maker – whilst facing the very real possibility of it negatively impacting their reputation or career if they voice their dissatisfaction. The result: inner turmoil.

Quite often, when a worker has made the conscious choice to complete a particular task set against their own beliefs and values, the resulting stress does not subsequently dissipate – but, instead, festers and becomes chronic. In these sorts of situations, organisations often see huge rises in worker turnover and sharply rising rates of staff being signed-off with mental health issues.

In other circumstances, workers rebel against 'the rules' because they truly believe that they are outmoded or over-cautious. This is regularly the case in compliance-heavy organisations where workers have decided that that processes are less dangerous than generally perceived – and that following all the rules represents something needlessly obstructive or inefficient. Typically, workers will begin to believe that since they are working on the shop-floor on a daily basis their judgement calls are more in touch with reality of the situation than the ideas of individuals that rarely visit their work area.

Though many progressive organisations put considerable time and resource into reducing levels of cognitive dissonance, it is rarely easy. Changing internal values is tremendously difficult at best – and, even when it is possible, it is a difficult, drawn-out process that needs considerable persistence. However, some organisations have found it possible to use rational data and challenging techniques to change thoughts and behaviour – and to promote safe working. Often techniques using role-play have been successfully used to enable workers to get a different new perspective on the day-to-day operations of an organisation.

It is possible to reduce the effects of cognitive dissonance by encouraging workers to understand that the dissonant belief is false, and then allowing them to resolve the situation by removing the conflict themselves. Though this is obviously not easy, it is possible to reduce the value of the conflicting belief by challenging a worker's assumptions with a number of thought-provoking questions that prompt them to reflect on their opinions or ideas. Naturally, this will depend on how deep-rooted the belief is. For example, it would be considerably more difficult to persuade an HR Manager to dismiss a worker for no good reason, than it would be to convince an assembly worker to wear gloves whilst operating machinery. Clearly, the first example involves a matter of principle, which is going to be significantly more difficult to change.

In order to encourage workers to re-evaluate their beliefs, it's often helpful to offer them substantive, factual information that aids them in changing their minds. Giving them evidence-based information that can be demonstrably backed up by credible sources will help – for example, proving that certain behaviours are dangerous through independent sources, such as statistics, may help empower them to make informed decisions and reach appropriate conclusions on their own terms.

Dissonance with incentives

Over the years, there have been many different ideas for attaining better safety standards within organisations, including the idea of offering workers incentives in order to behave safely. Known as the Induced Compliance Model, this approach was first outlined in 1959 in 'Cognitive Consequences of Forced Compliance', by psychologists Leon Festinger and Merrill Carlsmith.

As part of a social experiment, university students were asked to spend an hour doing a succession of dull, repetitive tasks – such as turning pegs in screws at different intervals. The tasks were designed to be tedious and induce strong, negative reactions with the subjects – which, from all reports, they were very successful at doing.

Once the subjects had completed their dull tasks, they were then asked to convince another 'subject' – actually a stooge – to take part in the same experiment by telling them it was really fun and interesting.

There were two groups of subjects: some were bribed with 20–dollar bills and others bribed with one-dollar bills. However, when both sets of subjects attempted to convince the 'new subject' that the boring tasks were actually really fun and interesting, they experienced dissonance: "The task was boring – but I'm lying so that this person will now do it."

As the two groups were given different-sized bribes to display a certain type of behaviour (i.e. to lie) they resolved their dissonance in different ways. The subjects that had been paid 20 dollars found it easier to rationalise their action: they had, after all, been paid a reasonable sum, which in turn made the lying more palatable.

However, the subjects that had been paid just one dollar found it much more difficult to rationalise their behaviour in the same way. In order to justify their behaviour they began to claim that they had done so because they actually believed the task was fun!

Both sets had rationalised their counter-attitudinal behaviour in markedly different ways, but when the second group couldn't reasonably rationalise what they had done in terms of a financial inducement, their inner conflict was resolved instead by changing their attitudes toward the boring tasks.

Festinger and Carlsmith determined from the experiment that in terms of 'induced compliance' the less justification provided for performing the counter-attitudinal behaviour, the more likely it was for an individual to have an attitude change.

Another reason why large incentives are unhelpful when used speculatively to improve safety performance, is that the rewards quickly become the only reason that workers value safety.

Giving workers substantial rewards in return for behaving safely tends to induce under-reporting of incidents and accidents in order to improve the chances of being rewarded. Additionally, when the reward is not forthcoming – or taken away – it causes workers to feel punished – and, consequently, respond by reacting negatively towards safety protocols.

Perhaps the best kind of inducements that can be used in the workplace are small and symbolic – and used in conjunction with rational coaching techniques which use questions, statistics and illustrations in the aim of getting workers to internalise promoted values.

A word on attitudes, beliefs and values

Managers impressing upon workers the need to for them to 'buck up their ideas' or improve their attitude is rarely helpful. Research by psychologists Joseph Tiffin and Ernest J. McCormick in *Attitude and Motivation* shows that more than 80 per cent of the respondents they interviewed had rated themselves as 'above average' when it came to attitude. It seems that like having 'a good sense of humour', having 'a good attitude' is just one of those things that most people believe they are imbued with. Though this is often palpably untrue to those around them, it's clear that many people may lack the self-awareness to see that they are largely deficient in these areas.

Tiffin and McCormick describe 'attitude' as: "the frame of reference that influences the individual's views or opinions on various topics and situations, and influences their behaviour". This is definition is widely accepted – as is their assertion that attitudes are largely comprised of beliefs and values.

Beliefs, considered to be based on the knowledge gained by empirical means, can vary greatly in their importance and influence. As a result, their resistance to change is also variable. For instance, an individual's belief in God is highly influential, and can be highly influential across a spectrum of other beliefs held by the individual.

On the other hand, values are far more intractable. They are comprised of cultural, ethical and social codes which serve to justify an individual's behaviour and are extremely difficult, if not impossible, to change. Therefore, it's essential for organisations to actively engage with their workers and find out what the prevalent values are – rather than simply sounding off with a set of arbitrary 'values' picked simply because they might sound good or likely to be supported *en masse* but in a fairly perfunctory way. It is crucial to ensure the corporate values are representative of the organisation and align with the values of the workforce – otherwise, performance will suffer.

Good safety boils down to good understanding, so solid research is paramount as a determining factor. To influence good behaviour in the workplace, instead of simply adding further levels of compliance or assiduously ticking things off checklists, it would make sense to spend some time to look deeper and discover what antecedents within the organisation are effective in creating safe behaviour and explore the consequences that inspire certain actions.

Rather than rigid rules and the threat of punitive measures, creating safety within organisations must be founded on the generation of appropriate attitudes and beliefs across the workforce. It needs to be something that is not coerced, but natural – and, ideally, and in time, comes without thinking.

A considered focus on behaviour is essential to the creation of a strong safety culture. It has been said that 'behaviour breeds behaviour'. It's certainly true that behaviour – both good *and* bad – becomes viral and sets the tone. Inside every organisation, each and every worker shares a fundamental interconnectedness with what goes on. Though it is not always obvious – it is, nevertheless, true. As a result, everything that is done – or *not done* – contributes to the safety culture of an organisation.

THE NAKED TRUTH: BEHAVIOUR

Much of what is written about safety behaviour (and indeed culture) focuses on academic and psychological aspects, so in this chapter we began by exploring some of the pertinent research around adult learning theory and the use of negative and positive reinforcement.

Behaviour-Based Safety has become a core component of safety improvement programmes for many organisations, and from its origins in Skinner's laboratory we know that in order to affect behaviour in a particular way, we must consider not just the activators (or antecedents) but also pay attention to the consequences too.

As human beings we are destined to make mistakes from time to time. By considering human error we learn that these mistakes can occur through lapses of attention, as active or latent failures or through cognitive dissonance. We've also considered why people may deliberately do the wrong thing and explored the types and impact of violations in the workplace.

Cultivating a deeper sense of understanding and engagement within an organisation can help mitigate the risk of error and improve the chances of sustained positive behaviours, whilst practical tools and techniques, such as using nudges, incentives and induced compliance can help in tasks where there's a risk of people tuning out due to repetition, habit or boredom.

The definition of culture as 'the way we do things around here' is often considered too simplistic – and, consequently, some people prefer to emphasise more complex pseudo-psychological alternatives. It is however hoped that this chapter has encouraged you not to overthink behaviour and to strip back your approach to how you and your organisation influence people.

The naked truth is that creating safety – and *safe behaviours* – within any organisation is more a question of addressing attitudes and beliefs. Instead of adding further levels of compliance or putting more safety officers on the shop-floor to watch workers, it's more a matter of thinking about how it might be possible to establish activators that encourage the right kinds of behaviours – and then providing desirable consequences that reinforce that behaviour. It sounds simple – and it can be.

GET NAKED: BEHAVIOUR

1. For the next ten minutes, as you go about your everyday life, think about what activators you encounter – and think about how the ABC model is at play in the world around you.
2. When you have considered the activators – ask yourself which ones routinely work on you. What consequences have a powerful effect? Do you routinely use the 'correct' bathroom in a bar or restaurant? Why? Does that work more or less effectively than the sign at the ATM advising you to protect your PIN? If so, why?
3. Choose an area in your workplace and spend a few moments identifying the activators for safety. How effective are they? If they are not effective, why not? Is the 'You are Entering a Hazardous Area' sign ineffective because it is obscured by a sheet asking people to sign-up to a new departmental five-a–side football team? Or simply because the implied consequence is too obscure? Which of these activators require some improvement now?
4. Consider the consequences that reinforce the activators. Are they desirable? How might your organisation create more safety by building more positive incentives for safety?

5. Review the incentives currently provided for safety in your organisation. Perhaps you reward teams for hitting targets on near miss reporting, achieving accident rate targets or completing BBS observations. Are you providing adequate acknowledgement and encouragement or creating induced compliance? What would happen if you removed the incentives?
6. What nudges can you think of beyond the workplace? In this chapter we've mentioned speed cameras, speed limit signs and warnings on cigarette packets, but which others can you find in everyday life? Can you identify three nudges that you might be able to bring back and try in the workplace to improve safety?

11 Leadership

The phenomenon of leadership has been central to the studies and observations of scholars, philosophers and military tacticians for thousands of years. Whether reading the ancient Athenian tragedies of Euripides from the fifth century BC, Plato's Republic *(428 BC), Sun Tzu's* The Art of War *(400–328 BC) or Machiavelli's* The Prince *(1513–1514), the reader quickly finds himself immersed in the art and science of leadership. Underscoring the importance of these early writings, such key texts continue to be found on the shelves of bookstores around the world. Testament to their present-day relevance, though, these mighty books are now located typically not in the 'Classics' section –but, more popularly, within the ambit of 'Business'.*

The global competitive landscape has changed significantly in recent times, forcing organisations of all shapes and sizes to respond, with disturbing regularity, to a gamut of changes thrust upon them from every angle. Accordingly, the need for clear and competent leadership has never been more urgent.

So, has leadership changed perceptibly since the days of Euripides and Plato? Are Sun Tzu's tactics relevant today? Are Machiavelli's principles still at play? What have we learned from the classics, and where will the future of leadership take us? This chapter reveals what lies behind the leader, exploring the traits, skills and abilities necessary to provide effective, impactful organisational leadership in the modern world.

Leadership is widely regarded as central to organisational success, but there is little agreement about what 'leadership' actually is, let al.one what traits, styles or techniques are best employed. Mired by myriad contingent factors that allegedly affect both its emergence and its efficacy, defining leadership has generated great debate over the years – particularly so in the task of distinguishing between leadership and management. Indeed, it would be fair to say that there may now be as many different definitions of leadership as there are people who have attempted to define it.

For a moment though, let's place the centuries of debate to one side, and take a helicopter view. One of the first classical organisational theorists, Mary Parker Follett (1868–1933), described management as 'getting things done by people'. Certainly a simple definition, but it helpfully underpins the tactical, transactional nature of the role. By way of contrast, Louis Pondy suggested that leadership should provide a clear sense of purpose of "what we are doing, and why we are doing it" indicating a much more strategic function.

Business guru Peter Drucker aimed to avoid the pigeon-holing that comes so often with attempts at definition of either 'management' or 'leadership' in isolation, offering that "management is doing things right, leadership is doing the right things". In Drucker's usual style, his explanation is certainly catchy, but does not really move us closer toward understanding.

In considering the *Anatomy of a Leader* Genevieve Carlowski (1994) suggests that whilst management is driven by the mind, leadership is powered by the soul. Esoteric perhaps, but a deeper dive reveals her logic (see Boxes 11.1 and 11.2).

BOX 11.1 MANAGEMENT

Descriptors:	Stabilizing, rational, consulting, persistent, problem-solving, tough-minded, analytical, structured, deliberate, authoritative.
Purpose:	Solves problems.
Process:	Directed, hierarchical, instructive process.
Force:	Uses the 'power of position' to drive action.

BOX 11.2 LEADERSHIP

Descriptors:	Visionary, passionate, creative, flexible, inspiring; innovative, courageous, imaginative, experimental.
Purpose:	Initiates and drives change to attain goals.
Process:	Reciprocal process which occurs among, within and through people.
Force:	Uses the 'power of the person' to influence others to act.

Carlowski's elemental analysis points North, quite literally, primarily noting the importance of having vision to succeed and creating a future for the organisation. As leadership guru John Kotter (1996) seeks to explain: "Leadership defines what the future should look like, aligns people with that vision, and inspires them to make it happen." Kotter points to the role of the leader as the compass, setting the direction, or perhaps the guide, showing the way forward. The actual 'creation', or action, of course, is done not necessarily by the leader, but by those around him who have been aligned with the path ahead and enabled to contribute.

So, taking our lead from Drucker and moving beyond the early thoughts of Parker Follett and Pondy into the twenty-first century, consolidating the numerous attempts at definition, today we can understand that management is about the application of problem-solving and structure in order to bring about order and stability, whereas leadership is the utility of vision and influence to bring about change through the actions of others.

The purpose and role of a leader

Even though many attempts have been made in popularising definitions of leadership and its core components, research has – with alarming frequency over the last six decades – continued to evidence that despite the volume of leadership models and theories, leaders continue to lack a clear understanding of their role – whether in times of organisational change, or during 'business as usual'.

Regarded by many as the world's leading authority on leadership, John Adair asserts that it's time to cut through the confusion and offers an elegant view that the purpose of the leader is to:

- achieve the common task *by*
- building and maintaining the team *and*
- motivating and developing the individual.

A succinct explanation, and helpful as a first step in our understanding that leadership is all about working through others. Theory is all very well, but how does this look in practice?

Traditionally, the approach has been to view the leader as the lynch-pin of all things, and so, taking this leader-centric perspective, let's explore the idea of purpose now and identify the six key functions of a leader.

1 **Clarity** – the importance of vision
How can we lead if we do not know the destination? Business leadership has much in common with military tactics, just like in Sun Tzu's *Art of War*, it's vital to move to the highest point to take a long-range view of the destination. Or as we say today, 'get up onto the balcony' and 'see the big picture'. Articulating the vision – and the reasons why we are striving to reach it – is the primary function of the leader.

2 **Engagement** – people power
In their excellent book *The Leadership Challenge*, James Kouzes and Barry Posner (2008) advocate that the purpose of leadership is to inspire the sharing of the organisation's vision by involving and empowering people. Leadership should establish a high sense of urgency in management and employee groups and provide clarity to help form coalitions and partnerships for action by role modelling the desired behaviours to build trust and faith. Encouraging people to constructively challenge the *status quo* builds a sense of shared responsibility, encourages teamwork and enhances engagement.
3 **Direction** – the success GPS
If people are the engine of the organisation and can power it forward, the leader then is the GPS or SatNav system, utilising her internal map and compass to provide structure, order and show the clear route to achieving the vision through the attainment of goals and objectives. Strategic thinking, planning and direction build the roadmap to success.
4 **Utility** – fulfilling the desire for meaning
Beyond vision and direction, the leader must create meaning for those around him by clearly, confidently and credibly articulating the vision so that it aligns people and inspires action right across the entire organisation. Without this, the organisation is reduced to a list of confusing, incompatible and time-consuming projects and activities all moving in disparate directions. Enabling others to understand how they contribute to the achievement of the vision is key. Relating parts to the whole satisfies our inner desires for meaning, and helps us make sense of the change we find ourselves in.
5 **Empathy** – the comfort blanket
People are governed first by their emotions, then by their intelligence. In a productive article in *Harvard Business Review* Heifetz and Laurie (1997) remind us of the importance of helping those around us to 'feel good,' and suggest that the most effective leaders hold a softer side to them, where demonstrations of empathy boost morale, instil confidence and help to build strong 'partnership working'.
6 **Lubrication** – keeping things moving today and tomorrow
Role of leadership is to hone, harness and utilise a range of interpersonal skills and organisational skills to identify objectives, navigate the route and steer the organisation to strategic success by binding groups together and enabling them to perform. In this sense, the task of the leader is one of keeping things running smoothly, effectively oiling the corporate cogs of the organisation.

Looking to the future, John Adair adds that leaders are responsible for 'releasing the corporate spirit' by identifying and developing the future leaders of tomorrow.

In sum, leadership is about having the right people, in the right place, at the right time, enabling them to do the right thing to progress in the right direction.

A brief overview of leadership theory

There is so much good academic research available on the topic of leadership theory that replication of it here in this book would be unwieldy. Synthesis, for similar rationale, would also prove beyond the scope of a single chapter. Instead, an overview of the predominant theories is offered as an introduction to encourage further exploration as your interest may be inclined. So, in our whistle-stop tour, we will find that leadership theory is typically rooted in one of three main areas, reflecting either:

- leadership **styles**;
- specific personality **traits**; or
- particular **contexts** or **situations**

Leadership Trait Theory

Popularised in the 1930s–1940s, Leadership Trait Theory revolves around the notion that there are specific personal characteristics which differentiate leaders from non-leaders. In simple terms: you either have it, or you don't; only if you possess the relevant personality trait, can you be a leader.

Certainly, we can observe that some great leaders – for example Bonaparte, Churchill, Carnegie and Ford – seemed to share particular traits such as *ambition*, *assertiveness*, *decisiveness*, *drive*, *persistence* and *self-assurance*.

Over the last five decades, Leadership Trait Theory has grown in research interest, and the traits of *adaptability*, *initiative*, *perceptiveness*, *tolerance of stress*, *intelligence*, *will*, *optimism*, *desire for learning* and *self-development* have all been added to the growing list of adjectives describing the traits of successful leaders. Even within the last few years, the capacity to *transform*, to *collaborate*, and an ability to be *charismatic* have found their way onto the hit parade, suggesting that not one single trait is necessary, but that successful leaders now depend upon a set of psychological traits to be effective. Whilst this is all very well, even the most cursory of consideration would develop a list spanning tens of traits, illustrating not a detailed picture, but more a 'paint-by-numbers' approach, where only those traits omitted from the list become conspicuous – by their absence.

Despite great interest and voluminous studies, the basic tenets of Leadership Trait Theory have never been truly substantiated. Even Ralph Stogdill (1974), the originator of the theory, began to doubt the simplicity of the approach:

> A person does not become a leader by virtue of the possession of some combination of traits . . . The pattern of personal characteristics of the leader must bear some relevant relationships to the characteristics, activities, and goals of the followers.

Stogdill's observation is crucial – possession of certain attributes is no guarantee of successful or effective leadership. Like the old English proverb 'clothes do not

REFLECTION POINT

Think for a moment of some of the greatest leaders you have known – what distinguishes them from others?

- Is it their ability to set the direction of the team, department or organisation?
- A sense of surety in their own actions?
- Their way of empowering others to action?
- An ability to make rapid decisions?
- Their power in giving clear instructions and orders?
- What specific traits have you observed that make these leaders stand out?

make the man,' neither do traits make the leader. Stogdill stops short in his self-criticism, but essentially points to the isolated and short-sighted approach adopted by trait theory as it evolved: specific traits will only be effective when utilised in certain contexts.

Heroes, Gods and the potency of people

In a superb little book aimed at boosting business performance John Fenton (1990) suggests that "leaders stand out by being different. They question assumption and are suspicious of tradition. They seek out truth and make decisions based on fact, not prejudice".

Throughout the ages, we've witnessed many leaders who have stood out by being different – Alexander the Great, Oliver Cromwell, Abraham Lincoln, Nelson Mandela and many more spring to mind.

These 'Great Men' reinforced classical leadership studies with the affirmation that certain individuals are bestowed with specific 'powers' at birth, and would naturally rise to the top and able to effect action with speed and certainty. Returning to those earlier words of Parker Follett, these people could 'get things done'.

Action by such worthies reeked of masculinity, individuality, authority and superior intelligence. People were typically viewed not as thinking, feeling humans, but more as a mechanical resource, effectively 'plugged in' to tools, plant and equipment to generate the required output. Despite this utilitarian approach, these people were not just leaders; they became regarded as heroes.[1]

And when we allocate leadership trait tags like medals of honour to this select few, we shift the concept of leaders as heroes one step further, following Plato's rationale for his 'Philosopher Kings', by imbuing those once mere mortals with sufficient potency to transform them into Gods.

Modern society continues to add fuel to the fire of the notion of God-like leaders. Business pages gleefully declare that Steve Jobs turned Apple around; Bill Gates brings computing to every home; Howard Schultz revolutionises a cup of hot coffee; Jeff Bezos defines internet shopping. What, all by themselves? Really?

Attributing God-like qualities to people does not guarantee God-like results. Whilst an important role, the leader may be highly exaggerated in terms of their actual personal contribution. Sure, it's easy to believe in the omni-potency of a single top leader to give us something to focus on, but isn't this delusion? As Richard Dawkins in his seminal book *The God Delusion* (1999) starkly offers: "There is something infantile in the presumption that somebody else has a responsibility to give (your) life meaning and point."

In contrast, leaders can be demonised too. Recall the global banking crisis, as one man, apparently single-handedly collapsed one of the world's biggest financial institutions. And why shouldn't people vent their frustrations on leaders? After all, they provide something clear upon which we can focus our terror, our disappointment and disgust. But does success or failure really come down to single individuals?

Leadership Trait Theory – a tentative conclusion

There are scant few particular traits that hold sound scientific evidence of a consistent impact on leader effectiveness – *motivation*, *drive* and *self-confidence* are amongst the meagre handful that repeat themselves across most meta-analyses.[2] There is a danger that trait theory soon turns turtle of course: as the list grows, we must question whether these traits really are distributed at birth, or whether, like other behaviours and skills, they can be acquired through learning and honed by repeated utilisation.

Leadership Trait Theory may be falling out of favour with the academics, yet despite the lack of empirical evidence, it remains popular today with many organisations using psychometric assessment designed to reveal prescribed traits during senior recruitment and appointment processes. Is it time for a rethink? Is there a case for accepting that the leading edge is within the organisational sphere and its desire to distil research, rather than a reliance on academia to continue to build their lists? Perhaps so. But with almost 400 individual academic studies completed to date, there remains no definitive list of traits that leaders must possess in order to be successful.

Leadership Trait Theory, therefore, continues to evolve less into a credible, useful model of leadership, and more into an almanac of amiable adjectives.

Leadership style theories

Leaders *are* capable of influencing events, of that there can be no doubt. But the diversity of contextual factors, political, social, economic present challenge –

even to Gods. Can it be, then, that there are particular *styles* of leadership which prove effective?

In 1960 Douglas McGregor presented two theories on the role of leadership. McGregor's view is that leaders either seek to strongly manage workers to complete tasks ('Theory X'), or to consult widely with subordinates to build relationships and encourage action ('Theory Y'). Subsequently, style theories have followed the X and Y lead, and generally evolved with leadership behaviour considered from either *task-oriented* or *behaviour-oriented* perspective.

The good news is that wherever you lie on the task–behaviour spectrum, it's the right place for you. The exercise is only an opportunity for self-awareness, something we may not always have much time to stop and do. As McGregor's style theory has evolved, there has been some consensus on its elements, so let's break down each side of the spectrum and look at how task and behaviour styles come alive in the workplace

Task-oriented styles of leadership

Task-oriented leadership is transactional in nature, driven by rules, process, procedures and standards.

Leaders initiate plans and build structure to clarify the roles and responsibilities of subordinates in order to facilitate completion of tasks. This style is especially useful in the achievement of clear, simple, uncomplicated tasks. Typically, task-oriented leaders are directive or pacesetting in style.

- **Directive** – Uses hierarchical power to tell people what to do and how to do it, expecting quick compliance with commands. There's little room for autonomy and tight leader-subordinate monitoring, supervision and control is exercised. Directive leaders rely on corrective (negative) feedback ('you're doing it wrong') and motivation is offered through a 'stick' rather than 'carrot' approach – noncompliance with directions will usually lead to negative consequence. The directive style has long been associated with traditional approaches to safety or risk management leadership – for example, how many times have you heard phrases like: 'We must do it like this or we'll get a notice/penalty/fine/have an accident'? Directive leaders often harbour intrinsic fear of failure or catastrophe, believing that the best way to survive is through clear prescription.
- **Pacesetting** – these leaders prefer to lead by example, setting high standards for themselves and their subordinates. Striving for excellence, the pacesetting leader expects that those around him understand the rationale behind the behaviours and actions being modelled. Reluctant to delegate priority tasks or those deemed 'complex', these leaders believe that such important or complicated actions require the absolutely highest levels of contribution. Rewards for performance are highly visual, with accolades, prizes and public decoration common methods of recognising positive contributions from the team. Pacesetters are usually unsympathetic to poor performance, though.

If their high standards are not met, responsibility is removed from individuals as quickly as it is given. For pacesetting leaders, it's a jungle out there, only the strongest and fittest will survive.

Fastidious task-oriented styles of leadership may be perceived by some as restrictive – stifling creativity and preventing the attainment of full potential through their controlling, compliance-driven nature.

Task-oriented leaders typically utilise an exchange-based definition of the employment contract – you give something, you get something. Subordinate motivation is generated through traditional 'carrot and stick' styles of reward and punishment. Whilst the task-oriented delivery style is unambiguous – and this may be beneficial in certain environments – it may serve to dampen constructive challenge and could eventually lead to subservient leader-subordinate relationships.

Behaviour-oriented styles of leadership

Behaviour-oriented leaders are big on relationships. Less focused on the formality of getting things done these leaders believe in the notion 'you get what you give'.

Optimistic, trusting and collaborative, the typical behaviour-oriented leader likes to be surrounded by his subordinates. For them, it's all about the people. Four types of leadership prevail:

- **Affiliative** – a people-focused approach, centred on mutual respect and trust. Believing that morale is the key influence on team performance the affiliative leader works on creating positive team interactions and building and maintaining a harmonious team climate. Affiliative types can quickly identify and respond to the needs and concerns of their subordinates, and actively seek out opportunities to provide constructive, often excessively positive feedback. They avoid performance-related confrontation and may find it difficult to offer negative feedback.
- **Visionary** – Taking a big picture perspective, painting clear visions for success, visionary leaders are keen to draw employee opinion and thought. Selling the vision is paramount for these leaders, so they take time to explain 'why' actions are necessary to drive long-term success. Articulation of the vision provides the necessary direction to inspire the team and build commitment.
- **Participative** – These individuals prefer to see leadership as a team game, striving for consensus of opinion. Participative leaders consult broadly across the stakeholder group, convening meetings to facilitate the sharing of best practices, co-creation of ideas and collective decision-making. In a bid to build employee engagement and commitment to the team, these leaders prefer to trust their subordinates to set their own direction in line with the broader organisational goals.

- **Coaching** – A popular approach to leadership over the last two decades, coaches recognise the importance and value of individual skills. They focus on enhancing the development of their subordinates by helping individuals to identify their own unique strengths and their potential weak spots and encourage the setting of professional development goals. As a leader, their primary focus is on creating a shared understanding and agreement of the purpose and scope of task roles.

Whilst easy to understand, and, for many people, quite palatable, modern style theories serve to underline the traditional view, percolated by the likes of Euripides and Sun Tzu, that leadership is an individual concept, in the sense that leaders differentiate themselves from others by some form of inherent (or learned) behaviour or style.

Potentially certain styles can bring dividend to particular issues, but they may cause a failure to consider the surrounding *context* of the situation. For example, how would an affiliative leader respond to dramatic organisational turbulence requiring mass resource cuts for example? Would the desire for inclusion and harmony hamper the ability to make decisions on staffing? In the same setting, might a leader with a preference for a directive or pacesetting style find themselves capable to execute the staffing cuts, but lacking the empathy to do so with sensitivity to the situation?

Where task-oriented leadership styles may be criticised from their narrow perspective on the work activity, behaviour-oriented styles are censured for their potential to allow leadership to drift away from the task at hand, and become absorbed in their relationships and quest for harmony. Such critique forms the extremes of a 'pendulum swing', upon which the other styles may be set.

The turbulence and uncertainty evident in most organisations today, in the wake of recession, during periods of instability, or global expansion, holds strong potential to impact upon and influence the leadership style of individuals. Remaining fixed in one particular style of leadership is not only insufficient but also inappropriate; our concern for productivity and for people must be *interdependent*, rather than mutually exclusive. Beyond this duality, a clear understanding of the context in which leadership is required becomes crucially important. Accordingly, we must go beyond an *'Either/Or'* model for choosing an appropriate leadership style and raise our gaze to include our surroundings in the organisational environment. A balanced approach – between task, behaviour and *situation* – is vital.

Leadership contingency theories

Contingency theories offer a natural step forward from style theories in that they *do* consider the context in which leadership is required. The golden thread is the assumption that for any given situation there will be one identifiable 'best' leadership style. For leaders with a penchant for tailoring approach, this perspective may be alluring. Over the years, three particular models have come to the fore and remain in widespread use across many organisations.

Fiedler's favourableness model

This model builds on the style theories by extending the point that leaders are motivated to act based on either their natural predisposition for accomplishment of tasks, or their desire to build and maintain relationships with others. Whilst the leader may seek to strike a balance between the two; flexing his approach as the pendulum swings, in times of stress or pressure the dominant preference will re-emerge and pull the leader back magnetically to his predisposition.

In such times, the efficacy of the leader in orchestrating group performance or success will depend not only on their preferred motivation, but also the degree of 'favourableness' to be found in the situation at hand.

According to Fiedler this 'favourableness' is influenced by three contingent factors:

1 **Leader–subordinate relations** – whether the leader is accepted and supported by the subordinates;
2 **Task structure** – the clarity of definition, structural composition, procedures and goals of a given task;
3 **Environmental influence** – the degree to which the situation enables the leader to exert hierarchical power and influence over the group.

Each element is measured on a given scale. High levels of each of the factors will yield the most favourable situation, with low levels generating 'unfavourable' or discomforting situations for the leader.

Fiedler's research identified that leaders motivated by relationships were most effective in reasonably favourable situations such as when working relationships are positive; tasks required to be performed are clearly structured and understood; and the leader holds strong hierarchical or 'position' power. Task-motivated leaders on the other hand are effective at either end of the range of favourableness.

Fiedler's belief was that the natural disposition of the leader anchors him to a particular preference, which in turn affects his ability to lead in favourable or

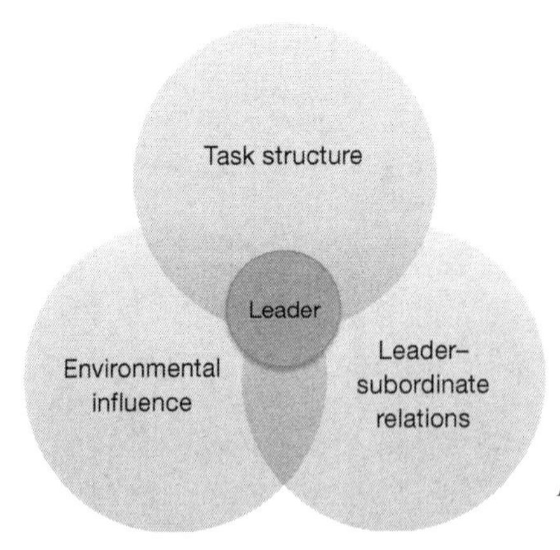

Figure 11.1 Favourable factors model

REFLECTION POINT: THE PERILS OF PRINCELINESS

In contrast to many other books in circulation at that time, which advocated the importance of kindness, virtue, mercy, generosity and peace if one desired to be a Prince, Machiavelli refuted this simplicity.

A strong reminder within his 'practical guide to leadership'[3] was that from time to time we must face choices between two unpalatable options. It's not possible to expect easy choices between darkness and light to always come your way, or to simply rely on gut-feeling to decide on what's right or wrong.

Instead, Machiavelli advocates looking at each situation as a unique opportunity, recognising the merits and demerits as they stand, and taking action based on robust assessment of these, distinguishing fact and evidence from perception and supposition.

Machiavelli understood the importance of contextualisation – even when faced with difficult choices. The true test of a leader for him was the ability to embrace the darkness, chaos and confusion, to see through it and take the right course of action.

unfavourable situations. The challenge for leaders is in finding their own personal 'sweet spot' between task, behaviour and situation. Fiedler suggests that this is perhaps not as easy as we might wish – and that it may be easier for leaders to change their situation – in order to gain degrees of favourableness – than to attempt to change their in-built beliefs and style. However, in the real world of modern business, perhaps we don't always have the luxury of just walking away from the less favourable tasks.

Path–Goal theory

In 'Path–Goal' theory, the leader's role is to identify and communicate the appropriate 'path' to be followed in order to reach organisational 'goals'.

The theory takes account of the conflicting interplay between the leader's preference for task- and person-oriented behaviours (as our pendulum diagram above illustrates), and the follower's levels of expectation and satisfaction. As the leader clears the path this provides inducement to allow the follower to move forward. This is done through offering reward for achievement, removing the obstacles that hinder or prevent achievement, and by providing the necessary support along the way. In this way, the theory advocates that the motivation of followers is engaged and this in turn increases the likelihood of attaining the organisational goals.

In its original form, the theory suggests that leaders select the most appropriate style for the situation. Four principal styles are offered:

- **Directive leadership** – ground rules are established to provide a solid framework in which to act and then specific instructions are given to subordinates to follow.
- **Supportive leadership** – the leader is sensitive to the needs of individuals and groups of subordinates, striving to maintain the strength of relationships.
- **Participative leadership** – wide consultation within the group before arriving at consensus-formed decisions.
- **Achievement-oriented leadership** – leverages the confidence built by and within the group, challenging goals are set and high performance encouraged by all subordinates.

In ambiguous situations or environments with considerable change, clear benefit and value may be derived from the tighter prescription of directive leadership. In stressful or difficult situations, supportive styles of leadership have been shown to improve levels of job satisfaction, reduced turnover, whilst reducing the potential for conflict. As the following true story illustrates, a blend of styles may at times be necessary.

REFLECTION POINT: BE SILENT OR BE KILLED

Wednesday 26th November 2008 had been a tough day in the office and I was looking forward to getting home and taking my dog for a walk in the hills.

I was driving home from work when my phone rang. Pulling over onto the shoulder, I noticed that it was the office calling – perhaps I'd left something behind.

"Andrew, we have a situation here," began my colleague, before going on to explain that a large number of employees from our business were staying in a certain hotel in Mumbai, India, which had been taken under siege by a cadre of armed gunmen. Could I get back to the office as quickly as possible?

Over the next two days our crisis management team worked tirelessly to systematically locate and expatriate our colleagues from India to their homes around the globe.

One employee, however, was difficult to find. A single colour photograph of Roger Hunt remained stuck to the wall of our temporary control room. Where was he?

In time, we did locate Roger; locked in his room on the 14th floor of the Oberoi hotel. With gunmen in the hotel lobby, and working through the floors of the hotel, Roger was frozen by fear, convinced he would not leave the hotel alive.

I made contact with him at first via email; his Blackberry device was his only lifeline to the outside world. After the reassurance of knowing where he was I began to take a more directive approach in my communication style with Roger as he explains in his book describing the event:

> Andrew now took on a dominant role in keeping me focused. And alive. His trail of cajoling emails gave me something fresh to think about. Andrew's persistence and ability to keep me motivated also showed the calibre of so many of my colleagues to rally round and work as a team.
>
> Andrew pestered and pestered. Every 15 minutes there was another message in the Inbox. At 4.15am he urged me to "email to acknowledge receipt of this message" and to "hang in there, buddy". At 4.43am, he fired another: "URGENT – READ IMMEDIATELY".
>
> He informed me that things were beginning to happen and gave me some essential instructions: do not barricade the door; have nothing in your hands; comply immediately with directions given. He went on . . . after another 15 minutes. . . with his 'Golden Rules'.
>
> He was still sitting in the control room with other colleagues – they had been there for nearly two full days, what wonderful devotion! At 5.56am, he pinged yet another: "Hello mate, sorry to bother you again, just want to make sure you're thinking of my 'Rules'! Stick in there, we'll both be needing a beer when you get home!"
>
> The emails continued in this way, then Andrew played a cute game with me; he said I'd get a pint of beer for every rule I sent back correctly. Never one to give up on free pints I replied: "Don't barricade door. Have nothing in my hands. Follow every instruction."
>
> Andrew responded immediately: "Don't know whether to be smiling or not, that just cost me three pints!!! Good work – you're a hero, stick in, hold tight."
>
> The Blackberry was my lifesaver. Andrew was doggedly persistent; the email exchange with him was keeping my mind off the gunfire outside my room. Another blast shook me. Andrew reassured me: "I understand that the explosion was a stun grenade as police are clearing the floors. Hang in there, buddy, stay strong and safe and stick to the Rules'." I repeated the 'Rules' to myself again: DO NOT BARRICADE THE DOOR. HAVE NOTHING IN YOUR HANDS. COMPLY IMMEDIATELY WITH DIRECTIONS GIVEN. Then Andrew said something I cherished:
>
> "I will not leave until you are safe." It was a kind thing to say.
>
> I had been told clearly what to do. There was no quibble about that. All my life I've known the importance of following instructions. I trusted Andrew Sharman with my life. He had given me 'The Rules' – I repeated them back to him.
>
> . . . then I heard the violent crash. "Open the door! Open the door!" came the voice with a heavy Asian accent. So, what did I do? I'd played the game with Andrew over and over again. For goodness sake, he owed me a few cold beers on this score . . .

On the morning of Friday 28th November 2008, Roger Hunt was evacuated from room 1478 in the Oberoi Hotel in Mumbai by a privately contracted military force. After 42 hours in a smoke-filled room, with gunshots, grenades and explosions peppering his ears, he couldn't believe that he was finally free – and safe.

Roger returned to Scotland, and our reunion was an emotional experience for both of us. We shed our tears as we bear-hugged each other. And then went straight to the bar – I had beers to buy for my colleague.[4]

Situational Leadership model

Paul Hersey and Ken Blanchard have themselves become heroes of modern-day leadership for their ability to distil the complexities of the subject. Their view on Situational Leadership offers a model that has become increasingly popular in recent years. The model expands the logic behind 'Path–Goal' theory and similarly proposes that there are four basic approaches to leadership: Telling, Selling, Participating and Delegating.

The key difference between Situational Leadership and 'Path–Goal' is the deeper consideration of the actual circumstance in which the leader finds himself. Prior to selecting a particular approach, the leader must first make an assessment of 'situational maturity'.

Whilst this may sound over-formal, in practice this is an internalised mental process calculating the overall maturity level by considering the psychological maturity of subordinates (their levels of commitment self-confidence, and their ability and readiness to embrace responsibility) and the job maturity (the degree of skill and technical competence held by subordinates).

- **Delegating** – High Psychological Maturity and High Job Maturity – responsibility for decision-making and implementation is given by the leader to the subordinates
- **Participating** – Moderate Psychological Maturity and Moderate–High Job Maturity – leader facilitates the sharing of ideas and decision-making within the group
- **Selling** – Low–Moderate Psychological Maturity and Low Job Maturity – leader explains his decisions and provides opportunity for group discussion to enhance clarity
- **Telling** – High Psychological Maturity and Low Job Maturity – leader issues specific instruction and maintains close supervision of subordinates

If subordinates are confident and motivated, and hold a reasonable degree of trust and readiness to follow the leader, a 'delegating' approach can be effective.

If they are less secure, or unwilling, they may respond more effectively to a stronger, 'telling' approach. As the levels of situational maturity increase, the leader can then adjust his leadership approach.

Whilst, on one hand, situational maturity-based leadership allows a clear approach to be taken, its generalized view does not fully account or allow for the variation of maturity levels within a group of subordinates.

REFLECTION POINT – WINSTON CHURCHILL

At any other time, Churchill could have been regarded as a failure. By 1935, the prevailing public view was that he had had his time. His best days in politics were behind him.

Despite his lofty aspirations, he never quite hit his own targets. Criticised for his opportunistic, outspoken, brusque and often leftfield style, he was destined for an early retirement of fine cognac and Cuban cigars.

Then, the Second World War came along. Churchill stepped up brilliantly to the task of Prime Minister, displaying decisive leadership and an innate ability to bring people together – in a strong, steadfast and ultimately resilient way.

Churchill's resounding success at leading Great Britain marks him as one of the most brilliant leaders of modern times and underlines the importance of not just ability and drive to lead, but also the crucial element of getting the timing just right.

Tentative conclusions on contingency theories

Contingency theories move us forward from academic conjecture to the real world of complex situations and changing work environments. They help us to underline that leadership isn't about recognising your own preferred style and playing that out in every situation, and remind us that there is a range of approaches that can be chosen and employed by a leader depending upon the situation he finds himself in. The theories typically acknowledge the fast lane of leadership by reflecting that, as human beings, we all have preferences for undertaking 'favourable' or pleasant activities, hold natural biases and enjoy the retreat into comfort zones from time to time.

Contingency theories, like much of leadership theory generally, may be traced back in time – for example, Plato, in his discourse on law and order, *The Republic*, pointed to the importance of situational awareness in shaping a leader: "A true pilot must, of necessity, pay attention to the seasons, the heavens, the stars, the winds, and everything proper to the craft if he is really to rule a ship."

Having a raised awareness of the composite factors and the range of approaches that we may choose to employ is, at the least, beneficial, and reminds us – especially in times of turbulence – that leaders must not only recognise, but also adapt themselves to the present situation in order to effectively utilise and leverage their inputs whether responding to, coordinating or leading change.

Several practical models have evolved to help us 'pay attention to the seasons'; their clear, logical approaches have been welcomed with open arms by leaders around the world. Despite decades of research on contingency theories, the conclusion must be drawn that we must still force ourselves to choose (and use) a certain style or type of leadership.

Models such as 'Path–Goal' and 'Situational Leadership' have allowed progress by distilling the types of leadership we might employ, thus making the choosing process a little easier, and have even de-dramatised the language from the empirical academic to the everyday pragmatic (Tell, Sell, Participate, Delegate). Perhaps this is why, despite the scarcity of hard evidence of their effectiveness, the popularity of contingency theories continues to grow. They are models that we can all understand, and find ourselves using. Even so, like trait theory, we still remain left without confirmation of any universal 'best style' that may be utilised. Our task of identifying the ultimate key to leadership success remains incomplete.

The romance of leadership

Over the last century, academic research has rigorously explored the realms of leadership, often taking its steer from military, political or other spheres. Whilst the output has undoubtedly been voluminous, the mass has made it difficult to identify the true essence of leadership. The challenge of identifying what makes a leader is now akin to finding a needle in a haystack. Or, rather, with thought to the volume of theoretical mass, more like finding a needle lost between three very large haystacks.

Trait Theory began life as a worthy cause – however, over time, has become over-burdened by the steady stream of adjectives loaded upon its shoulders. Being personable, smart, strong-willed and ambitious are no longer sufficient – with more than 200 different traits now all posited as 'essential' for leadership the theory has become lost within itself. Through its mass collection of personality traits, the theory has naturally, though ironically, evolved into a signpost for clarity, pointing to the fact that not one particular aspect makes the leader, but, instead, a combination of traits which arguably then become a certain style or preferred approach.

The overloaded Trait Theory encourages us to view style theories not only as easier to comprehend, but also more relevant to the concept of individual personality. Style theories shed light on how each of us hold natural preferences with regard to how we work, either keen to progress and complete tasks, or with a bias for building congruent relationships with those around us.

This element of Style Theory will resonate for most of us who are sufficiently self-aware to understand our comfort zones and preferred ways of working. However, dependency upon a particular style is a bit like placing all of our eggs in one basket. In a previous world of work, where industry stereotyping was possible, and where task operations were reasonably consistent, developing a fixed style may have brought solid returns. Today, with the advent of globalisation, geographical and cultural diversities restrict the efficacy of adopting a singular style and implore us to look more broadly at the context in which we find ourselves.

This significant shift in how work is done highlights the importance of contextualising leadership with the situation, and it's here that contingency theories bring their value. By continuing the theme of personal preference (or 'favourableness') this perspective acknowledges the import of individual approach. Path–Goal Theory reminds us of two of the key functions of a leader – to remove the roadblocks that prevent others achieving their goals; and providing motivation to drive action forward. Clearly logical in its approach, it's no surprise that this theory still holds water with so many organisations today. Contingency leadership theory also delivers us a crash course in the assessment of maturity. Hersey and Blanchard (1988) strip back the jargon from the theory and show us that in order to 'get things done' we simply need to consider the situational maturity and then adjust our style to suit the situation.

Each of these theoretical approaches offers some assistance to us as we follow the evolution of thought. Each provides us with useful *perspectives* on leadership, yet stop short of our goal of truly understanding what leadership *is*. A common thread amongst the perspectives discussed in this chapter is that each recognises the value subordinate contribution. From a common-sense point of view, most people would agree that success is a team game, so if we accept that results are borne out of collective action, why do we continue to persist in linking success firmly to the actions of the leader?

The answer lies in that fact that with so much uncertainty surrounding the anatomy, mechanics and science of leadership – what it is, how it works, and why – it becomes far easier to ignore the greyness, and in our search for simplicity quickly conclude that leaders are automatically regarded as the cause or source of leadership.

> In the absence of direct, unambiguous information that would allow one rationally to infer the locus of causality, the romanticised conception of leadership permits us to be more comfortable in associating leaders with events and outcomes to which they can be plausibly linked.
>
> (Meindle et al., 1985)

In practice, the process is incredibly straightforward – a mental heuristic or 'shortcut' is actively taken when we ascribe leaders with control and responsibility. Our brains, without us knowing, perform the following logic sequence:

> She is responsible; therefore, she is a leader.
> She is a leader; therefore, she gets the results.

But the trouble with this heuristic – and indeed the problem with each of the dominant leadership theories – is that they are myopic: they only see the world from the leader's viewpoint, and discount the perspective of subordinates, followers, peers or other stakeholders. Worse, they ignore the broader contextual forces which influence and impact upon organisational performance. What is created, then, is a kind of 'Paradise Island'; a land where the leader rules – *because* she is the leader. When we attach further labels – subordinate, follower, manager – and try to construct convenient pigeon-holed definitions, we only further the disconnect from the actual reality of the activity.

Disconnected leadership and the need for a new approach

There has been far too much focus on the leader in the past, ignoring the broader context and the contribution of others. This over-glorification has caused not just the rise, but also the fall of leaders, and the art of leadership itself. A disconnect between our view of leadership and the world we live and work in is becoming glaringly apparent.

Yet, organisations are still predominantly structured like the ancient pyramids, with one leader at the top. Like the singular Pharaoh of bygone days, most organisations have one single CEO. Both Pharaoh and CEO have been regarded with the possession of exceptional abilities. Both 'great men'. But the leadership assumptions that have continued to prevail, are now, like the pyramids, thousands of years old. Just as we no longer choose to build new pyramids, we must move our thoughts on leadership into the present day.

In order to progress, we need to leave behind the notions of great men, heroes and gods of leadership. We must celebrate the complexities that are collaboration and contribution, and see the world of work through new eyes, to look beyond the theory. A new perspective is borne out of change like a phoenix rising from the ashes of the bonfire of vanities. 'Post-theoretical post-heroic leadership' is the way forward. Why? What has spurned this need for new style of leadership? There are three highly persuasive reasons:

1. The rapid and turbulent change experienced during and immediately following the global recession;
2. Feelings of discontent around the image of leaders as strong, masculine, authoritative, and numerous public outings of once glorified corporate leaders who brought downfall to their businesses or industry sectors. We must move beyond this Machiavellian manipulation; *and*
3. The ever-increasing challenges of managing workplace diversity across organisations which continue their quest for globalisation.

The complex environment of the modern organisation requires a different approach. Previous theoretical models have suffered from over-formality, and being too static, too rigid, to cope effectively with a fluid, dynamic fast-changing world. We must change. We now need a paradigm shift in leadership.

In safety and risk management we often use an Iceberg Model to illustrate the relationship between insured accident costs (the small tip of the iceberg) and those that are uninsured (the bulk below the waterline), we can apply the same logic to leadership: the leader is the tip, and the vast network of collaboration and support generated by managers, followers and others beneath the waterline. Without the bulk, the tip would never show above the waterline and reap the successes. Yet the success is contingent on a broad variety of knowledge, abilities and skills, that go beyond the scope of just one single person, and it's exactly this view that we need to carry with us as we explore new ways of leading.

Relationship theory – a transformational approach to leadership?

Reflecting the evolutionary process in both management and leadership, and recognising the importance of contribution from the entire workforce, relationship theories are concerned with the connections formed between leaders and subordinates. More intimate than the other previously discussed theoretical models, relationship or 'transformational' leadership revolves around the potency of leadership 'charisma' to motivate and inspire people to see the greater good and holistic importance of the task at hand, encouraging subordinates to go above and beyond what's expected of them. Old problems are reframed as new challenges to be solved collectively and creatively, encourages constructive challenge of the status quo.

Such theories aim to cultivate an atmosphere of positive change, fostering a sense of purpose by aligning people with the organisational vision, encouraging innovation, ownership and empowering them with the responsibility for the achievement of goals the leader builds trust, improves morale and enhances a sense of self-worth in their followers.

Is transformational leadership just a 'sexing up' of traditional leadership theory? After all, the power is still held by the hierarchical senior, who through his social interactions drives others to cause events. Certainly, these leaders are indeed focused on the performance of the group in pursuit of strategic success or performance improvement, but – due perhaps to higher ethical and moral personal standards – transformational leaders recognise that followers do hold diverse needs and concerns and so strive to support each and every person in the fulfilment of his or her own potential.

Whilst phenomenally popular and at face value an appealing move forward from the more traditional trait, style or contingency theories, there is little empirical evidence to confirm that transformational leadership actually does 'transform' individuals or groups in the workplace. Despite the lack of evidence, we may be seduced to herald the future of leadership through faith in such 'transformation', but there exists a very real risk – of falling back to outdated 'great men' beliefs – as transformational leaders too become elevated to the status of organisational gods.

But if we are cognisant of this risk, find a way to suspend any feelings of expectant awe for just a moment, dare to defy the academics, and feel brave enough move beyond an insatiable desire for hard-fact evidence if we look very carefully we may just notice that the *technique* of transformational leadership has much more to offer . . .

REFLECTION POINT – VARIATIONS ON A THEORY

As the London Philharmonic Orchestra approached their climax of Edward Elgar's *Enigma Variations*, I sat in my seat, stunned by the power of the performance. I counted 76 musicians on the stage, each with their own instrument of different shape and size, yet all playing in perfect harmony and alignment with each other. Wondering what could be the secret to this sweet success, my mind moved to the music. They all had a copy of the same musical piece right there in front of them, all they had to do was follow the 'instructions' that were written in front of them. And then my explanation vanished as more than half of the players laid down their instruments and sat, still as statues, for a moment.

The music continued to wash over me and my mind tried again to solve the riddle. Perhaps each musician has a specific set of instructions – a set for the violins, another set for the brass section, another again for the percussion . . . surely a better idea . . . but as the bass drum boomed its rebuke, I knew again I had missed the point.

It was only as I began to give up on the puzzle and return myself to the ecstasy of the sound that I noticed it. First the frizz of grey hair, next the flash of a hand, then both hands thrusting skywards, a short stick twirling in the air as if stirring an imaginary cocktail . . .

This musical miracle was not just a case of following instructions, receiving the necessary training, having a degree of psychological maturity or even being told what to do, but indeed it was *caused* by the collective collaboration of a large group of participants, all of whom were following the non-verbal gestures of one man. His twirling baton, his nuanced hand movements and his subtle but striking facial expressions – each influenced not just the pace, position and power of the musicians, but also their commitment to the piece. Without saying a word, this diminutive sexagenarian man in spectacles perched there at the front of the stage, with his back to the audience, became invisible as the audience focused on the players – but yet here he was, with a gentle-as-a-mouse approach, not just controlling, but *leading* an entire orchestra – with constant adjustment and refinement – not just to play great music, rather to enable and empower them to give their all and deliver the finest rendition of Elgar that they have ever performed in their lives.

Effective transformational leaders rely not on their words to lead, but through the consistent deployment of symbolic gestures and behaviours, underpinned by a strong set of values – including empowerment, engagement, inclusion and communication – which *cause them* to lead others by example. It is this 'power of gesture' that can differentiate transformational leadership from other approaches in recent years. It is this understanding of the magnitude of gesture that sets British conductor Hilary Davan Wetton apart from so many of his contemporaries. Yet, it is this very element of gesturing that has been so grossly under-appreciated in organisational leadership. The impact of clear communication from non-verbal means is vitally significant, now, more than ever, at a point where cultures collide and we find less time available each-and-every day for dialogue, but, perversely, there are more people than ever before watching our every move.

The 'Great Man' traits of masculinity, authority and superior intelligence – none were observed in Davan Wetton's conducting. With his back turned to his stakeholders, the paying audience, he was 'getting things done'. No heroics, no god delusions, just 'doing the right things' to allow the collection contribution to ensure that the whole performance was greater than the sum of its *77* parts. Now, isn't *that* leadership?

The dawn of post-heroicism – moving from transformational to distributed leadership

Davan Wetton exemplifies the concept of 'post-heroic leadership'. At the end of the performance, he is the first to recognise the contribution of the musicians, warmly bowing to the players before inviting the audience to join him in sharing gratitude. His sense of humility, taking his own bow to the audience only on the cue of one of his performers, demonstrates that he places himself not upon a pedestal above it all, but instead, there *with* the musicians, supporting them, showing them, guiding them, leading them.

It is time to look beyond the academic rigour, to lean away from the traditional research and models, and seek leadership lessons beyond organisational behaviouralism. Process-driven leadership with its linear, masculine style of one-task-at-a–time cannot survive in a constantly shifting world. Adaptive approaches to leadership, with allocation of styles and models for given circumstance may offer a framework for leaders to lean on, and there will almost always be a time for the prescription of the directive or 'telling' style, whilst in other moments a hands-off delegation will suffice.

To make a step change though, we must review *our perspective* on leadership. The old view positioned the leader as a 'supertanker', powerful, awe-inspiring, determinedly focused on reaching port. Common business metaphors strike hard at the idea of super tankers or ocean-liners, the time it takes them getting up to speed, the slowness of turning such a large ship to make adjustment to its course. We know from Hollywood movies and media frenzy that when a big ship founders it quickly spells the end. We've simply got to see leadership in a new light.

History tells us that with the right leadership, even failing super tankers can be supported and complete disaster averted, but as our knowledge and understanding of the complexities of business has developed so to we must now reposition organisational leadership. The 'Command and Control' tactic for the super tanker needs to be replaced by individual decision-making under advised direction.

REFLECTION POINT: A TALE OF TWO SHIPS

The nautical metaphor acts as a useful precursor to a tale of two extremes that underlines the importance of leadership visibility in galvanising calm and commitment.

At 11:40pm on 14th April 1912, Captain Edward John Smith was woken from his bed by one of his staff to be advised that their ship, HMS *Titanic*, had collided with an iceberg.

Over the next three hours, Captain Smith led his crew from the bridge of the ship, coordinating emergency repairs, summoning rescue support, assembly of passengers and launching of lifeboats. For nearly the entire time, Smith remained at the helm of his boat, leaving only to aid physically in the handling of children into the lifeboats. The records of perhaps the most well-known shipping disaster of all time reveal that at 2:10am, as the *Titanic* lurched forward for the last time, the ships Second Officer saw Captain Smith walk calmly back to the bridge after seeing off the last of the lifeboats. Seconds later Smith was engulfed by the waves. Certainly, one of the greatest tragedies at sea, the sinking of the *Titanic* claimed more than 1,500 lives, but without the solid leadership of Smith, many more would have perished.

Now consider in contrast the actions of Francesco Schettini, Captain of the passenger cruise liner *Costa Concordia* which struck rocks just a few hundred miles off the shore of Italy on 13th January 2012.

The Captain deviated from the route scheduled for the trip, in order to perform an unsanctioned close proximity 'sail past' the island of Giglio. Despite the calm seas, the ship was steered too close to landfall and its hull was ripped open by the reef at 9:15pm. An hour later, at 10:20pm, the Captain tried to calm fears by advising that the ship had merely suffered an electrical failure, causing the electrical power to drop out, but within minutes passengers had donned lifejackets and some leapt for their lives into the ocean. Seeing no cause for alarm, Captain Schettini went to his quarters and ordered dinner.

At 10:42pm, Schettini decided to report the incident to the Italian Port Authorities, who promptly dispatched rescue support. For some passengers, this was already too late. An hour after taking his meal, the Captain abandoned his ship and travelled to the mainland on one of the rescue boats. Over the next few hours, Captain Schettini was ordered back to his boat to take control of the disaster, however he did not return. At 3:05am, on 14 January, 600 passengers were evacuated from the *Costa Concordia*, and, finally at just before 5am, the evacuation was deemed complete. In the subsequent investigation – in which Schettini was charged with manslaughter – 32 people lost their lives. The "shortcomings in the actions of the Captain" were found to be the primary cause of the accident, and the actions of Francesco Schettini following the impact with the reef were considered "tragically inadequate".

We must see the task as ships moving together, cohesively in a fleet according to plan. Where the super tanker maintains only one single focus – arriving safely at port and discharging its load, the distributed fleet can maintain several foci, distributing or sharing the leadership responsibility. In a complex work environment, especially those spread geographically across the globe, could strategic aims be better served *collectively*?

As organisational hierarchies become flattened and 'matrix structures' become the norm, there is now a pressing need for leadership influence at all levels within the organisation. Why? Because flattened structures encourage proximity to customers – whether external, or internal to the organisation, intimacy is quite literally forced upon us. What could corporate life be like if lengthy communication channels were avoided? If employees were positively encouraged to make decisions, take the lead and drive progress? Speed of response brings added value, but beyond that, is an absolute imperative in the fast-paced world in which we live and work.

Of course, such self-organisation requires trust, stability and experience. I'm not advocating a leaderless society, there's a reason why car manufacturers have not yet launched fully automatic vehicles, but I do suggest that a shared approach, or distribution of leadership could prove useful.

One for all, or all for one?

Before our minds dream idly about organisations without leaders, like cars without drivers, we must ask the question – *should* leadership be the domain of one person, or should it be distributed throughout a team?

It's impossible for such a broad competency profile, traits, style and the capability to always consistently select the right approach for every situation to exist in one person. As DeMarco and Lister point out in their excellent book *Peopleware*:

> On the best teams, different individuals provide occasional leadership, taking charge in areas where they have the particular strengths. No-one is the permanent leader because that person would then cease to be a peer and the team interaction would begin to break down.

Certainly a 'revolving' leader position would avoid the isolation of one person holding charge, but how would we continue to ensure focus, strategic alignment, continuity?

Distributed leadership does not mean the wholesale removal of *all* leaders. If we are to move past the heroics of leadership, we need to move our focus from one mere individual-as-leader to a collective dynamic. So, let's stop our searching for the best possible leader and instead find ways to cultivate the emergence of distributed leadership.

There remains a need for a figurehead to point the way. A leader who is flexible, understanding and able to articulate vision and delegate tasks but also someone who has the emotional intelligence to be able to know when to pass the baton to others for their leadership as and when required. So, where do we start? Well, perhaps rather ironically, it starts with a single person, a single leader. It starts right here with you.

Authentic leadership

The notion of authenticity is gaining in popularity as the new panacea for leadership, and it's not entirely surprising. If leaders do not remain true to themselves, and hide behind a mask of perfection, slickly acting out the orders of the day, it's highly unlikely that subordinates will trust them and bring themselves to work day after day to follow their lead.

The word authentic comes from the Latin 'autos' for 'self' and the word 'hentes' meaning 'being'. In simple terms then, authentic leadership is about being who we are. Despite centuries of conjecture as to what makes the best leader, it's ironic that the latest conclusion is that what makes the best leader is . . . you!

So, it is indeed time for us to move away from the ideas of leadership gods. Yes, there have been some terrific organisational leaders who have made great impact; perhaps their biographies do offer enlightenment on how they gained their successes. But instead of striving to become more like Steve Jobs, Jack Welch, C.K. Prahalad or Gary Hamel, why not take a leaf out of Richard Branson's book and be *exactly who we are*? Our leadership style, skills and abilities *emerge from our experience*, not from our emulation of others. This copycat idealism, promulgated by the constant drip-feed of 'secrets of success' and 'keys to achievement' broadcasts on the world's social media networks serves only to undermine our own authenticity, rather than develop it.

Authenticity is a crucial element for leadership to embrace as one of the building blocks for effective engagement. Authenticity is shaped by how a leader's values influence the words he says, how he conveys them, the actions he makes and how he makes them. These elements are like the spokes of a wheel, with trust as the hub that holds them all together. Leadership, too, is just like the wheel, a virtuous circle in that when you lead effectively, people follow willingly. This circle of leadership brings responsibility towards others, a responsibility that is either given or taken away based on one factor: trust.

As the model below shows, it's a circle: giving trust to others, building their trust in us. It's also about trusting ourselves – to take the right action, to make the right call, to do the right thing. It's easy to use the word, and it may be common to observe leaders emphasise the importance of developing mutual trust within a relationship, but oftentimes these same leaders forget to consider their own levels of trustworthiness and how this is perceived by those around them. It's vital to remember here that we cannot *direct* trust to occur, *because* we are a leader, but that trust is the *response* to our words or actions, following judgement by our peers, stakeholders and others.

Building trustworthiness is about demonstrating that you, as a leader, are trustworthy. The British philosopher Onora O'Neill suggests that there are three core values that effectively help us gain the confidence and trust of those around us:

1 Competence
2 Honesty
3 Reliability.

Competence is of course, crucial. Over the years the media has identified incompetence from all walks of life and pitched it into the public area under the auspices of 'justice' much in the same way as Plato's *The Republic* operated. Competence is not simply about having the necessary skill to perform the task, but understanding our own strengths and weaknesses, knowing when to draw the line upon reaching the limits and being able to delegate the action to one more suited.

Honesty – some of the sagest advice comes from Samuel Clemens, better known perhaps as Mark Twain. Ever seeking to chronicle right from wrong Twain tells us "When in doubt, tell the truth." Why tell the truth? For two reasons, firstly: "It will confound your enemies and astound your friends." Second, "If you tell the truth, you don't have to remember anything." Unassuming council from one of the most beloved writers in modern history.

Reliability – Reliability is doing what you said you'd do, when you said you'd do it. It's not just consistency, but the dependability of that consistency.

O'Neill's suggestions on the core values for leadership provide a powerful foundation stone against which we can consider our own authenticity. We must remain mindful that possessing these values is not enough in itself. The way we demonstrate them is crucial: the words and tone we use when we speak, the actions we make, and the behaviours we employ – each and everything

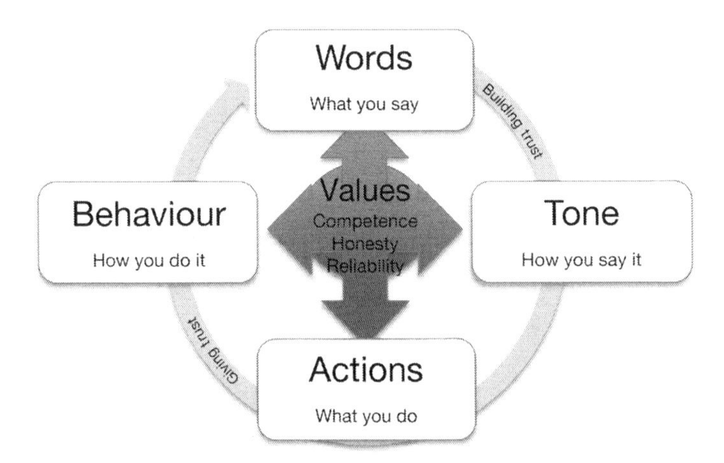

Figure 11.2 The circle of trust

we do reveals how seriously we live the core values. The most successful leaders know that building and maintaining trust hinges upon establishing trustworthiness in not just their own behaviour and actions, but in the behaviour and actions of everyone involved. So where does it all start? With the authentic leader, of course.

Authentic leadership

The former Chairman and CEO of Medtronic, Bill George, knows a thing or two about authenticity. In leading the company, George spearheaded such dynamic change in the way the business was operated, that across an 18–year period the Revenue Annual Growth Rate was an impressive 18 per cent; Earnings per Share annual growth rate an astonishing 22 per cent; and Market Capitalisation 32 per cent. How did this happen?

George is resolute on the needs of modern business. He makes it abundantly clear that "executives running corporations into the ground in search of personal gain" appear to get hired for their apparent charisma and ability to ramp up stock price rather than their ability to lead. Charm and glamour are not prerequisites for good leadership: "We do not need celebrities to lead our companies." Whilst some boardrooms may have red carpets, they are not catwalks. Instead, George points to the power of character, integrity, commitment. In short, authentic leaders. Leaders who "have the courage to build their companies to meet the needs of all their stakeholders, and who recognise the importance of their service" (George, 2003).

George advocates that authentic leaders demonstrate five primary qualities:

1 Understand their purpose as a leader;
2 Practice solid values;
3 Lead with the heart;
4 Establish connected relationships;
5 Demonstrate self-discipline.

Emotional intelligence

Effective authentic leadership depends crucially on having sufficient emotional intelligence to recognise that there is always room for improvement. Regard the writings of any of the greatest business writers and gurus of our time – Welch, Adair, Kotter, Collins, Goleman and many more – and the same message appears again and again: leadership skills must be constantly practiced, honed and developed.

It's fair to say that technical skills are necessary to get us into the leadership game, but it's emotional intelligence that keeps us in play. Leaders with high levels of emotional intelligence recognise that their every interaction influences the moods, emotions and responses of those around them. Self-awareness is key, as is the degree of self-management under stress. Emotional intelligence has previously been tagged as part of the 'soft skills' required by leaders (and managers) but recently this aspect of leadership has risen to the top with even legendary business guru Jack Welch positing it into the 'hard-edged business skills' bracket.

REFLECTION POINT: AUTHENTIC LEADERSHIP

Think of 5 leaders you respect and admire – they can be from any area of life or work; religion, politics, law, technology, academia, science, the arts. You may know them personally, or not. List their names here (there's no need to rank order them)

Leader	*What I admire*
1	
2	
3	
4	
5	

Looking at each in turn, reflect on why you admire them. It's more than likely that once you have moved your thinking past their achievements or accomplishments, you'll find your appreciation connected by an emotional element – the *how* they do what they do. This emotional element is the '*who*' they are. This is their authentic self.

Leadership or followership?

The resurgence of the 'Great Men' theories of leadership and the public celebration of 'corporate gods' credited with turning organisations around or revolutionising industries is starting to wear thin. Within the last few years, leadership has been over-glorified. As the world settles itself back into reality in the wake of global recession, where have the leaders gone? Arguably, they're still around, but maybe, just maybe, they have realised that there could be an alternate view.

'Followership' has been posited as 'The Next Big Thing' in leadership circles. Really? People have followed leaders for years. In religious sects, political parties, organisational groups, social conventions, in financial investments, fashion and even in the schoolyard playing tag. Followership is nothing new.

But followership *needs* a catalyst for action. Let's look at four examples.

Bradford

Fifty-six people lost their lives and around 300 others were seriously hurt on Saturday 11 May 1985 when a fire broke out in the spectator stands at the Valley Parade Football Stadium in Bradford, England. Examination of the live television broadcast footage reveals that despite the fire spreading rapidly, fans did not react immediately, choosing to remain to watch the fire and the football game. Why did these people stay where they were?

New York

On 11 September 2001, Marissa Panigrosso was at her desk on the 98th floor of the South Tower of the World Trade Center in New York. As the first aeroplane hit the North Tower, Marissa not only heard the impact, she felt it too, as the shockwaves rushed through her building and knocked her from her chair. As she stood to make her way to the emergency exit, Marissa noticed that many of her colleagues continued to go about their business; making calls, filing papers, some even entering a meeting room to begin dialogue. As she walked away from her desk, Marissa noticed that two of her colleagues remained at their desks chatting on their telephones. Despite the wave of panic that was beginning to sweep through the floor, the majority of people carried on with business as usual. Marissa recalls clearly those final moments of her colleagues: "I saw [them] on the phone . . . I remember leaving and [they] just didn't follow."

Geneva

With around 160 employees in the small building tucked on the outskirts of a major city, it was an average Tuesday. It was exactly 11:04am on 16 April 2013, when the fire alarm broke the morning routine. In the open-plan office on the first floor around 50 employees were busy with their work, diligent, even. Though the shrill of the bell was disruptive, they carried on with their accounts, transactions and data-basing. The managers and team leaders remained ensconced in their glass-fronted office cubicles, glued to their telephones, or busily tap-tap-tapping away at their keyboards. The clock reads 11:07am. With the exception of the piercing wail caused by the alarm, all looks normal. The occasional furtive glance from one employee to another, scanning the room to check for signs of new action, to confirm the need for panic. Others keep themselves low behind the screens dividing their work cubicles, seemingly afraid to pop their heads up. An Admin Clerk eventually stands up from her desk, walks briskly to the other side of the office in the direction of the exit, but stops short, at the photocopier. Looking through her team leader's glass wall, she sees him continuing to work, and returns smartly to her desk, facsimiles in her hand, and begins to stuff the envelopes waiting for her. At 11:14am the sound stops. Fire drill practice is over.

Glastonbury

Imagine you're at a concert in a park, sitting on the grass, basking in the sun, listening to some great music. Suddenly, a guy near you jumps up, flailing his arms and legs wildly, grinning like a Cheshire cat. For the first few moments you may watch in surprise, but then, as people around you start to get up and join this dancing man, copying his moves, you become fascinated – it seems that people just want to have fun like he is. More and more people join the crazy dance, falling in line behind him, until most of the park has assembled. Glancing around now you realise that you are in the minority; most of the people are now also flailing around with silly smiles on their faces. What do you do?[5]

Negative panic

Negative panic is the name given to the behavioural phenomenon that causes us to stop and remain still at times where our instinctive reaction should be to go, get out, escape. We wait, patiently, it seems, for corroboration of the emergency. We seek to decipher what's happening around us, waiting for smoke, flames or explicit instruction from a trusted source. We wait for someone to take the lead. We wait for someone to follow. For all of us, there's a tipping point – the moment where a sufficient proportion of those around us have engaged in a certain action – that forces us to follow. Where is yours?

To lead, to follow

We began this chapter by distinguishing between managers and leaders, and it seems appropriate to return to that point as we close this discussion. A manager is someone who is followed because they must be obeyed. A leader is someone who is followed by others of their own volition.

Followership can be a viable alternative to traditional perspectives on leadership. The ingredients in the mix are straightforward: someone with the courage to stand up and show what needs to be done; repeat the demonstration; as the first few followers join, provide encouragement to allow the first followers to feel safe – even if they are dancing crazily. The initial actor (or 'leader') is still there, but now the followers are the ones being followed, they've turned into 'secondary leaders' who in turn inspire others to act.

Followership depends on our ability to reach, connect and positively influence people beyond hierarchical structure, persuasive logic or intellectual reasoning, and this naturally must be the true aim of the leader. From a foundation of core values – perhaps including honesty, reliability and competence – the authentic leader envisions success and engages others to enable collective success.

Managers may be appointed based on time-proven skills, dedicated service and loyalty to the organisation. Whilst the leader may have less practical skills, it is her vision that unites people and urges them to follow. Without followers, the leader remains just a man or woman doing (or, arguably, not doing) their job. Going back to our earlier aphorism, *clothes do not make the man* – but followers certainly do make the leader.

THE NAKED TRUTH: LEADERSHIP

Distilling what constitutes effective leadership is not easy. However, one thing that leaders through the ages – whether Bill Gates or Alexander the Great; Bonaparte or Bezos; Cromwell, Churchill or Captain Smith – seem to have in common is that their leadership inspires others to take action.

Whilst the purpose of leadership (to lead others in the achievement of goals) may not have changed significantly, there has been a sharp shift in its context. Now, in the Knowledge Era, a broader set of leadership competencies is required as a mix of vision, authenticity, values-based relationship skills, are required to help leaders take the lead on focusing on global issues facing the organisation.

So, what do good leaders do? They clearly communicate the vision; set direction to orient people towards common objectives; they establish a sense of pace and urgency to drive action. They articulate purpose and meaning; and by building trust they encourage collaboration and the development of collective responsibility. They facilitate debate on issues and actions, instilling self-confidence to allow constructive challenge of the *status quo*. A good leader celebrates group and individual achievement, sharing sincere praise and recognition where it's truly warranted – and appreciated.

It's virtually impossible for a single individual to possess the depth of knowledge and entire breadth of competencies to consistently lead effectively in this fast-changing world. So, moving leadership from the directed function of a singular person (the traditional 'straight-line' leadership of old) to a more authentic approach enables others to give their best and take a share of responsibility. Smart leaders understand that trying to fit within the confined boundaries of traditional theory no longer resonates with business operations and aspirations in a rapidly-changing world, and constrains not just their own personality, but also impacts negatively on their personal and team effectiveness.

True leaders see beyond the detail, and bring clarity, depth and texture to articulate the vision and turn it into reality by building understanding and commitment amongst their followers.

A critical determinant of effective leadership is not how brilliant the strategy is, but how clearly followers understand it and are able to demonstrate the behaviours necessary to drive the organisation towards its aims.

The greatest challenge for leadership today is how to lead and inspire a team who you may have not met, and perhaps never will, in a broadly globalised, highly unpredictable world. Lead not as you have been led; but lead as you *should have* been led. Leadership is an art and a science, so develop your own leadership style which is true to yourself, strip it back, be more human.

GET NAKED: LEADERSHIP

1 Are you, as a leader, clear on what your role is in terms of 'providing leadership'? Which of the 'six key tasks of a leader' do you feel you need to focus on right now?
2 How comfortable are you with your position on the Task-Behaviour spectrum? What does your current role require? Where would you prefer to be?
3 Do you know a leader who appears able to consistently inspire action in people around them? How do they do this? What techniques, traits or approaches can you identify?
4 Thinking about your own team, what do you need to do in order to improve levels of Job Maturity and Psychological Maturity?
5 Transformational leadership is about empowering people to optimise results. What subtle gestures can you employ to help you engage, encourage, enable and enhance the work of those in your team?
6 Which of the following two statements do you most agree with:

   a I trust people until such time as they prove themselves to be untrustworthy
   b I don't trust people until such time as they prove themselves to be trustworthy

   What could the impact of your choice be when it comes to your relationships at work?
7 How does your view on trust affect your own trustworthiness? What opportunities are there for you to enhance your personal trustworthiness?
8 Can you think of a time where you may have hidden behind a leadership mask? What can you do to be the authentic leader that you already are?

Notes

1 Whilst the 'Great Men' theories of leadership began with a masculine perspective, subsequent writings have added female leaders such as Margaret Thatcher and Mother Theresa to the ambit.
2 See e.g. Kirkpatrick and Locke, 1991.
3 *The Prince* by Niccolo Machiavelli wasn't strictly a 'practical guide' – though Machiavelli may well have intended it to be – but so many lessons within its pages have resonance and validity for leaders of every era, in every industry.
4 Extracts taken from the book *Be Silent or Be Killed: The True Story of a Scottish Banker Under Siege in Mumbai's 9/11* by Roger Hunt, published by Corskie Press. The attacks in Mumbai in November 2008 claimed the lives of 164 people, and injured over 300 more.
5 Search for 'The Dancing Guy' on YouTube and you'll quickly locate one of the most fascinating studies in Followership in recent history. It's such a powerful little video clip that Harvard, TED and many others at the cutting edge of 'thought-leadership' have picked it up to distil its charm. When would you have joined the group and started dancing?

12 Leading with safety

In this chapter, we'll continue our discussion on leadership by exploring how safety can be led in the workplace. We begin by considering how expanding corporate values into a set of safety principles can provide a strong platform to lead from, and how these can in turn help us to manage expectations around safety leadership more broadly. A return to the notion of authentic leadership is included to help bind the elements of values, principles and purpose together.

This chapter will then dive into the relatively recent concept of felt leadership *to understand just what it is that makes this type of leadership quite so powerful when it comes to matters of safety. A model of felt leadership combing three interconnected pillars is presented to explore the impact of active leadership, the importance of worker involvement and the necessity of robust review.*

In this book, you will find a chapter discussing organisational leadership, and then you'll find this chapter here, on safety leadership. The inclusion of two separate chapters is deliberate. It's not to suggest that the two topics are discreet and exist in isolation to each other, but the contrary, to pull out the elements of leadership that are most relevant to leading with safety, and to explore these in more practical detail. At its most fundamental, safety leadership is not dissimilar to organisational leadership, and indeed much of the science behind the subject is borne out of the evolution of leadership as a discipline in its own right. But if we believe that workplace safety is more than just reducing accidents and injuries, safety leadership must be more than just hanging a vision on the wall and giving people safety glasses and boots.

The traditional approach to safety and risk management is task-oriented: the completion of a risk assessment, creation of a work procedure, delivery of training, supervision of new team members, investigation of an accident, review of risk assessment procedures, revision of procedures – and repeat.

Whilst, at first glance, this may remind you of the spiral of continuous improvement, I offer instead that the process is linear, rather than circular, because natural evolution has so efficiently streamlined our inherent ability to learn from our mistakes, that now we take benefit from each activity that occurs, and almost automatically derive an improved way forward. This has generated good progress for many organisations and it's common to monitor the 'success' of our approach – the culmination of the risk assessments, procedures, training and more – by plotting the (hopefully downward, improving) trend of accident figures over time, and perhaps flagging significant safety tasks as 'milestones' for action on the journey.

In modern organisations that are committed to safety, performance typically improves over time following the systematic introduction of advanced technology, the implementation of formalised policies and procedures, and then the advent of management systems that brings everything together. And then, with all the procedural, systems and technological work done, what comes next? The human factors side of safety. It's here that many organisations turn expectantly towards behaviour-based safety programmes and 'employee observations' in the belief that they offer a panacea. But many such programmes go straight to focusing on the behaviour of the workers, and miss the crux – leadership.

The sequential linear process of identification, pursuit and action of specific tasks has facilitated sure and steady progress in safety improvement. For many organisations, with an unwavering and consistent commitment to workplace safety, a 10–fold improvement in accident rates can take on average between 10 and 15 years. A successive, incremental rate of positive change can often be sufficient for some –however, eventually, this step-by-step, incremental change will slow, slow, slow and then finally draw to a halt. We reach the fabled 'plateau'. We have become 'good'. We have attained a rate of accidents that is lower than we've achieved before. We may even have better rates than our competitors, or find ourselves leading our industry sector. But is it enough? The modern answer of course, is 'no', it's not enough. We strive to further improve. In the days where the phrase 'zero accidents' is chanted like a mantra from the boardroom to the shop-floor, do we have the patience to wait 15 years?

In their report covering more than 21 countries in developed and emerging markets *Business Pulse: Exploring the dual perspectives of the top ten risks and opportunities in 2013 and beyond*, Management Consultants Ernst & Young show that health, safety and environmental risks remains at the top of the list of concerns for senior executives. The report states that organisations now have an: "intense focus on safety and environment risk preparedness and mitigation" and suggests that "in light of Corporate Social Responsibilities, economic challenges and regulatory pressures, it has become increasingly clear that managing these risks has become vital for long term sustainability". It would seem that for many organisations there's an urgency to act. With societal pressure for good corporate citizenship, and the ever-tightening grip of legislation how much longer *can* we wait for the safety performance we desire? More to the point, *why should we wait*? Is there a way to expedite progress on our journey to zero? What's the key to allaying boardroom concerns?

Safety leadership is the answer

Whether as a leader or as a risk management professional, you likely harbour a desire to establish a solid organisational culture that values and promotes safety from the boardroom right down to the factory floor. A culture which encourages personal responsibility and the engagement of employees at all levels. A culture that is borne out through everybody demonstrating the right behaviours all the time. This organisational culture can only be created by *leadership*.

Why is *safety leadership* so important, then? Because it's a catalyst. A catalyst for action. But more than this, safety leadership is a powerful catalyst for organisational cultural change. Culture is affected by two specific inputs from the leader – their *intention*, or 'what they say', and their *execution*, 'what they do'. In safety terms, a leader's *intentions* manifest not just as the words that leave their mouths, but also those words that inevitably hang from the walls too, the policy and mission statements, the safety vision and values. Their *executions* refer to how the leader demonstrates her commitment. This may be through her own behaviours – such as wearing the appropriate protective equipment, whether or not she holds the handrail on the stairway, uses her mobile telephone whilst driving – and also via the broader commitments such as the amount of resource invested in improving safety in the workplace. Safety *leadership* is the glue that binds together the standards, systems, processes and everyday actions of the individuals and groups within the organisation to create an effective route to safety excellence.

The evolution of safety leadership

In the early 1970s in Great Britain, when the government ordered a root and branch review of workplace safety laws, Lord Robens, in proposing a new legislative architecture for health and safety, was so convinced that the concept of safety leadership was so vitally important that it became formally enshrined in the modern foundations of UK safety law – The Health and Safety at Work Act.

Leadership as the foundation stone for workplace safety has remained a constant feature of modern law around the globe, and beyond the legislature too. Employers' organisations like the Confederation of British Industry or the International Organisation of Employers; Non-Governmental Organisations such as the World Health Organization and the International Labour Organization; and professional bodies including the Institute of Directors, the Institute of Leadership & Management, the American National Safety Council, and the world's largest organisation for safety and health practitioners, the Institution for Occupational Safety and Health, all advocate that solid leadership in safety is the critical first step. In fact, the Institute of Directors is so convinced of the urgent need for leadership to embrace health and safety that it has produced an excellent guidance booklet defining leadership actions for directors and board members entitled *Leading Health and Safety at Work*. The booklet opens with a clear statement that: "Protecting the health and safety of employees or members of the public who may affected by your activities is an essential part of risk management and must be led by the board."[1]

So, the commitment of the organisational leaders is a crucial key that either locks or unlocks the power to deliver success when it comes to workplace safety and risk management. Why should this be so? The rationale is, of course, remarkably facile. When something holds weight and importance at the highest level of the organisation, it is highly likely that action will be taken to achieve its respective goals.

Certainly, the leadership teams of most organisations today are *interested* in having good safety performance, and indeed many will have safety as a standing item on their boardroom agendas. But herein lies the rub: more often than not, this 'leadership commitment' to safety manifests most commonly just as that: 'interest'. Yet, the monthly high-level glance at the top-line metrics is scant review to ensure that Key Performance Indicators (KPIs) are tracking as the board has been advised they would. Once confirmation of agreeable data trends has been gained, the executives may be satisfied that they are leading safety efficiently and move right on to the next agenda item.

The learning paradox

What causes such cursory self-assurance? Why would executives happily accept data dashboards and promptly conclude that sufficient attention is being levied? Just as with our previous question, the answer is remarkably straightforward: organisational leaders rarely, or dare I even say *never*, receive any training in *how to lead on matters of safety*. Pull up the syllabus for any executive learning programme or MBA and the odds are stacked highly against it ever featuring workplace safety. Why should it be so? Well, there's the argument that safety is just simple common sense, right? Wrong.

Go back to those learning syllabi – chances are you'll notice that quality assurance, logistics and productivity are all in there in some shape or form. As they should be, after all these elements are inherent to the laws of business – whether supplying services or manufacturing products, it's essential that the leader understands how to get each of these right. Look back at the course outlines and you may notice that customer service is included too; after all, we want to make sure our customers are happy – now *that's* common sense. Look again and you'll find that human resources, performance management and communication skills are all there, indicating – perhaps rather ironically – that the sense required to work with and through people is not as common as we might hope.

Supportive leadership?

The lack of formal learning on safety spins us into a vicious cycle, which, whilst from the outside may look as if it's positive, when we look deeper we find that the action it encourages not only restricts progress but also gradually erodes the good practices in place.

Many organisations charge a board-level Director with the responsibility for safety. Frequently, this may be the HR leader ('because safety is all to do with people'), perhaps in some cases is may even be the most senior leader: the President or CEO ('because they're ultimately responsible for everything') or, as is rapidly becoming common, the Operations Leader ('because operations is all-encompassing'). This leader acts as a figurehead, lending their name and support to the workplace safety agenda. Often such leaders have climbed through organisational hierarchies, leaving behind their own personal early specialisations, and evolving into a generalist, holding the strings to a broad range of organisational disciplines or functions.

Leaders can find themselves in their senior positions after moving steadily up through the ranks, some years after starting out in a specific chosen field or discipline. As they progress, they may continue to excel at their chosen specialism, whilst delegating other elements to others when they don't have the confidence or competence to lead. Indeed, the concept of top tier generalised leadership governing mid-level specialization has become popular across many developed countries, and is now firmly embedded as the *style du jour.* For these leaders, providing personal support and delegating seems perfectly normal. Especially if they have not benefited personally from formal training in workplace safety and risk management it seems natural to pass the baton rather than risk looking incompetent.

Whilst a delegated approach may work efficiently for say logistics, production or quality assurance, in matters of safety things can quickly fall apart. Good leaders realise that building and maintaining a team of technical experts around themselves can complement their own skills and expertise – after all, we can't all be experts in everything, right? But if this occurs with safety it's not safety leadership – its leadership giving safety some support. In their work with large and often complex organisations, psychologists Robert Kegan and

Lisa Laskow-Lahey conclude that: "the leader must be more than a mere supporter," painting the details into their picture they suggest that

> we cannot succeed if the leader is only authorising our participation, if he or she is merely a sponsor of work being led by outsiders. We rely on the leaders we work with to be genuinely partners and when the resistance mounts, as it nearly always does, it is the leader . . . who must help the group renew its commitment to the journey.
>
> (Kegan and Laskow-Lahey, 2009)

Great leaders understand that it is *tasks* that are delegated, not *leadership.*

Let's just pause for moment and take a sense check. I'm not proposing that *certain aspects* of safety cannot be delegated. In fact, I would propose that there are many aspects of workplace safety and risk management that can indeed be more efficiently completed when delegated. Like in any other function, there will be those better suited to specific tasks or activities, but the visible demonstration of leadership interest and commitment, at the most senior level in the organisation is not one of these.

As we saw in the earlier chapter, the primary role of a leader is to lead. Not *to support.* Demonstrating leadership *support* for workplace safety is not only a sure-fire way to diminish its importance to the organisation, it's also self-contradictory. When leading safety is delegated to the technical specialist, the leader has essentially abdicated responsibility. Safety then rapidly loses its status as a business imperative and drifts into becoming: 'something that gets done around here when we have time . . .'

The purpose of safety leadership

So, what should the leader charged with responsibility for safety actually do? In the earlier chapter on organisational leadership, we discussed the purpose of a leader and identified six key functions. When we look back at these, we see that the first four are *vitally relevant* when it comes to leading with safety.

These high-level functions provide a framework for good leadership, whatever role is held within the organisation – however, at the same time, these primary functions are indeed those necessary to set the tone for safety leadership.

A distribution of safety leadership

With the absolute commitment and the utmost dedication of the 'official safety leader', dramatic improvements in workplace safety cannot be guaranteed. Even if this person does happen to be the CEO.

Traditionally, the leader at the top was viewed as responsible for the success (or failure) of the organisation. Whilst there can be no doubt that the leader holds great influence over the organisation, if we pick up one of the hundreds of autobiographic business leadership books from former business CEOs and look

Clarity All too often in safety and risk management we carry ourselves away with technical jargon, abbreviation and language choice. The safety leader must disentangle and translate; shining a bright light onto the vision, mission and objectives and explaining why safety is important to the organisation, and to her personally as a leader.

Engagement Clarity is the first step to engagement; without it, there is only confusion. Engagement is not the same as delegation, nor instruction, but the act of involving others on an emotional as well as a physical level, fostering real interest and gaining genuine commitment.

Direction As John F. Kennedy declared: "Efforts and courage are not enough without purpose and direction". It is insufficient to expect others to look to the vision and always see the next steps on the route forward. The safety leader must show the way.

Utility The leader must ensure that safety activities are relevant and practical to the tasks or operations at hand. Followers must be able to understand why such actions are required, feel able to undertake them and see the results of their implementation.

closely at the organisations that have gained outstanding success in recent years – whether science or technology, heavy manufacturing, consumer goods or service provision – we learn that each one has thrived not because of the overwhelming input of just one single person, but in fact the contribution of a *team* of leaders. This concept of 'sharing' leadership is not just relevant to safety and risk management, but in fact it is critically important to breaking through that performance plateau we looked at earlier in this chapter if we want to achieve and sustain levels of excellence in safety.

The CEO isn't off the hook just yet though – this 'distributed' leadership is a productive approach to leading with safety, but it must *begin* with the leader. It's imperative that the top leader has a genuine and clearly visible commitment to health and safety to set the example for the other leaders and crucially demonstrate that safety is still owned and delivered from the top.

Just like the conductor leads his orchestra. Try imagining the outcome if all the musicians were left to do their own thing without that overarching sense of direction from the conductor! Similarly, without leadership of safety from the top, there is no legitimacy given and quickly the approach falls into disarray. But when the top leader leads clearly this sets the tone for others to follow. Then, when each of the hierarchical organisational leaders are visibly, consistently articulating and delivering the vision *and* role–modelling the expected behaviours as part of their everyday activity something magical happens. Others don't just become followers; they become empowered to also lead in their own way with safety. Distributed leadership for safety in this way is a high-impact method for gaining real commitment, traction, engagement and results.

Who *is* the leader?

As legendary leadership author Stephen Covey pointed out, when most people think of 'leadership' they think of a position, and therefore may not consider themselves as 'leaders'. But this self-disqualification has no relevance when it comes to workplace safety and risk management. When I speak of 'safety leaders' I'm not referring to the technical expert, the person holding the 'Safety Director' job title. Instead, I am speaking of a much wider group. With *Naked Safety*, the safety leaders within the organisation are of course the CEO and his executives, but they also extend through the hierarchy to include the operational leaders, the directors, managers and the front-line supervisors. Because it is *these leaders* that are responsible for driving results within their respective areas and it is *these leaders* who constantly shape and influence the behaviour of the workforce around them on a daily basis through *their intentions* and *executions*. These leaders are each individually just like the orchestra conductor, *distributing* their leadership to influence their followers. Behind the leaders, just behind the stage curtain, stands the technical safety expert, composing the music from the background.

A distributed leadership goes hand in glove with a distribution of power across an organisation. This is not power in the traditional hierarchical perspective, for the flattened-out matrix structures in many companies today have helpfully removed much of that, but instead power in the sense that everyone is able to make good decisions based on an acceptance of their inherent ability to do so through clarity of understanding of their role. Recognising the power of responsibility in this way, the workplace suddenly becomes more vibrant, more stimulating, more challenging and offers more opportunity for developing a sense of self-worth and personal skill sets.

Leadership and personal values

So, where do we start? How do we 'distribute' safety leadership? Well, just like with the orchestra, safety leadership needs to be 'conducted'. So, we start right here – with *you*!

Each and every one of us holds deep within us a set of personal values that shapes our thoughts and guides our actions on a daily basis. Sometimes this happens consciously, more often than not it occurs without us noticing. Being clear on what our personal values are – and how they relate to the people we work with and the tasks we undertake – is key to maximising our potential as individuals, and crucial in leading effectively.

In the box overleaf you'll find a list of commonly held values. Quickly review the list and underline or circle the five most important values to you in your life right now. The values don't need to relate to work, or to safety, they're simply what's most relevant and important to you. Each of us holds a different set of values, unique to us, and these have been honed over our lifetime. For most people, personal values are relatively constant and don't change frequently, therefore work through this exercise briskly – you should find that you notice your values quite quickly.

Achievement
Accountability
Adventure
Aesthetics
Affiliation
Ambition
Authority
Authenticity
Autonomy
Challenge
Change
Collaboration
Competence
Cooperation
Creativity
Equality
Excitement
Fairness
Fame
Family
Fast pace
Flexibility
Freedom
Friendship
Fun
Harmony
Happiness
Health
Helping others
High earnings
Honesty
Independence
Integrity
Job satisfaction
Loyalty
Order
Participation
Performance
Personal development
Physical effort
Power
Precision work
Pressure
Quality
Recognition
Reliability
Responsibility
Reward
Safety
Security
Self-expression
Spirituality
Stability
Status
Structure
Thought leadership
Tranquillity
Trust
Variety
Wellbeing
Working alone
Working with others
Work–life balance

Extrinsic Values are the high-level aspects, often related to our ambitions and desires. As the name suggests, these values are 'beyond us' as individuals. Extrinsic values are useful in the sense that they motivate us to achieve a purpose – specific job title, promotion, power, prestige, financial gain, task completion and so on. (Extrinsic values connect well with the 'Task' end of the 'Task-Behaviour Spectrum' discussed in the chapter on Leadership). Extrinsic Values can help us to focus on business goals and objectives, targets and financial commitments.

Intrinsic Values are traditionally considered to lie at the heart of ethics and ethical behaviour. Revolving around such aspects as humanity, family or sense of moral duty to others, Intrinsic Values may also help us to achieve a purpose, but at the same time they *are* the purpose or reason we do things. They may lead us to an extrinsic benefit, but action is taken because it feels right to us as individuals.

REFLECTION POINT: PERSONAL VALUES

Turn the spotlight back on yourself. Without thinking too deeply, what are the most important things to you, right now, from a work perspective? Note your thoughts below as they occur, resist the temptation to fine-tune or edit as you go.

___

___

___

___

___

- Are there any surprises in your list? Do you feel that there's a balance between Extrinsic and Intrinsic Values, or do you have a natural bias towards Extrinsic or Intrinsic Values?
- How does your natural value set influence your personal safety leadership?
- Is your leadership style driven more by the notion of personal connection or task completion? How does this affect and influence those around you?
- How can you harness your personal values to positively influence action in safety?

Considering which values are most important to us as individuals is helpful in understanding our own personal motivation. Our personal values can be split into two perspectives, extrinsic, and intrinsic values.

Great safety leaders possess a deep and finely tuned sensitivity to their intrinsic values and strive to personally connect with individuals in a meaningful way, recognising their personal worth and contribution. Driven by intrinsic values the safety leader commits herself emotionally to the health, safety and wellbeing of those around them – even where those others are not personally known, or even acquainted to the leader, the sense of moral duty (in this case to keep people safe from harm) fuels their concern, commitment and, crucially, their actions.

Principles and the corporate compass

Our personal values shape our beliefs and approaches to work, govern the way we interact with others and influence the way we lead others. Over the last decade or so, values have become increasingly important to corporations too, and much of the academic research on corporate culture has emphasized the 'value of values' and underlined how much impact clear and concise meaningful values can have on forming, shaping and sustaining the culture of an organisation. Often, values act as a kind of 'corporate compass', pointing the way forward. In this way, the corporate values can help to identify the principles of how the organisation operates from a safety perspective, and this in turn will provide a solid structural framework to allow safety leadership to grow.[2]

Identifying principles for safety is not a complex task, especially for organisations that already have a clear view on their safety vision and values.[3] In the box below you'll find some of the more common principles that came out of our study. How do these compare to your own personal and organisational principles?

So, with a clear set of principles for action, what should we expect safety leaders to do? In my own recent research working with safety and business leaders from more than 100 blue-chip organisations around the globe we looked at how safety leadership is viewed by senior organisational leaders. In this study, whilst almost all the business leaders agreed that distributing safety leadership was a great idea, less than 20 per cent of the organisations involved had undertaken a process to identify what they expected from their safety leaders. Here are some suggestions that emerged from our study that may help to get your own thinking started.

SAFETY LEADERSHIP PRINCIPLES

- Safety is of equal importance to other business objectives such as production, quality, service delivery.
- People are at the heart of workplace safety.
- Management systems, policies and procedures must *enable* activity and support workforce needs.
- Everyone has a right to intervene in matters of safety.
- Be authentic – never hide behind policy, rules or hierarchy.
- Go beyond your own team.

SAFETY LEADERSHIP EXPECTATIONS

- Demonstrate genuine concern for safety.
- Set clear safety expectations.
- Demonstrate personal safety behaviours.
- Never walk by – act immediately an issue is identified.
- Recognise and praise positive safety behaviours.
- Identify poor safety conditions and act to improve.
- Strive for continuous safety improvement.
- Provide (and accept) constructive feedback.

A return to authenticity

By setting principles and expectations we're off to a great start. Whilst it's easy to say that safety is all about people, it's even easier in practice to miss this point, and instead, inadvertently, turn safety into a 'numbers game', ranked alongside other performance data such as the numbers of widgets made, customer deliveries shipped on time or defect rates. At this point, it becomes exponentially more difficult to reposition safety as part of the organisational values framework. So, how do we *as individual leaders* avoid falling into this trap? We return to our discussion on authenticity.[4]

Authenticity is fast-becoming one of the most talked about attributes in modern business. Whether seasoned CEO or starting out on the career ladder, we're told we need to be more authentic. But being authentic can be a serious challenge. Allowing others to see us for who we *really* are may be a daunting prospect. We'll need to admit that we are *human*, and that as a result of this – yes, wait for it – *we sometimes make mistakes*. In the Knowledge Era, information is power. To err is fatal. So, how can we be more authentic given the high stakes and strong tide of resistance?

Spiritualist Andrew Cohen has a fantastic list that he advocates if we seek to be truly authentic as leaders. Cohen suggests that courage is the first step to authenticity, and then we must be willing to:

- **Live fearlessly** – live and work with the courage of your convictions
- **Act heroically** – take action for the benefit of the masses without the need for personal gain
- **Stand alone** – hold and articulate an opinion, standing up for the decisions we take
- **Have unfaltering commitment** – to your goals, beliefs and actions to achieve them
- **Live for a higher purpose** – Work tirelessly to help others achieve their goals and aspirations
- **Face everything and avoid nothing** – developing awareness and understanding in order to drive progress and achieve results, even in the face of adversity or challenge
- **Take unconditional responsibility** – being responsible for our actions, acting ethically
- **View with perspective** – regard your environment objectively, impersonally and without ego

Much of Cohen's list resonates well with safety leadership. Imagine the impact of your work if you lived fearlessly, facing everything with a 'can-do', 'no-holds barred' attitude, holding no pre-conceptions that would cause you to hesitate or shy away. If you gave your absolute commitment and assumed full responsibility for the safety of everyone around you, acting to develop practical ideas to ensure that people go home safely at the end of the day. What influence would this have on those around you? Would it not energise them, enthuse them, motivate them and inspire them?

Authentic leadership is not about ego and crafting the perfect professional image. It is about reconnecting yourself with your values, focusing on your purpose as a leader and being willing to act for what you believe in. Cohen suggests authenticity comes when we care

> so passionately about others also reaching that goal that we unhesitatingly sacrifice our own peace of mind, comfort, and security for them to succeed. It really means that we have no choice left anymore because we have realised without any doubt that from now on, it's up to us.

Felt leadership

Being 'authentic' is at the core of great safety leadership. That makes good sense, after all, safety, in essence is about caring for people, right? Authenticity is so glaringly obvious that it's easy to tell when a leader is paying lip service to safety, or just adding their name in support of safety projects in the workplace. We can just somehow *feel* it. Martin Luther King understood this well, and encouraged people to judge him by his *character*. This may seem like a subtle point, but it's crucial to our understanding; as this is how safety leaders (and arguably, all leaders) in turn are judged. The most useful measurement of character when it comes to safety leadership is not a Key Performance Indicator set by the leaders themselves, or even by the technical experts, but rather by those receiving the messages communicated by the leader – both in word and in deed: the followers.

As we conduct our work in accordance with the three core values of competence, honesty and reliability we build the 'circle of trust' that enables us to be 'who we are', and encourages us to utilise our personal values to lead from the heart with safety. Leading from the heart allows us to 'pin our colours to the mast' and helps others to see our character, to see just what we stand for. Think back to the exercise you completed in the Leadership chapter where you reflected on leaders who have inspired you. It was their *emotional intelligence*, their *character* that had you hooked on how they led. Recall your list of leaders from that exercise. When you think of these leaders again now, would you say that they led with the conviction of their values; that they led with their hearts?

In recent years, the world of workplace safety has come to learn and use the phrase 'felt leadership' after the DuPont company identified that this was a core component in its own improved safety performance. DuPont asserts that felt leadership begins with the leader *feeling* and *believing* in the core values of the organisation. The next step is to *act* in accordance with these values. It's at this point that many leaders often struggle. 'What should I do?' 'What do I say?' 'How do I do it?' may be common cries from within the organisation, but safety is not a mystical back art. Let's strip things back and look at exactly what felt leadership looks like in practice. Leaders who are *felt*:

- set a good example;
- know the operation;
- anticipate risks;
- discuss hazards;
- are alert for unsafe conditions;
- follow up;
- inspect often, intelligently;
- take effective corrective actions;
- investigate incidents and accidents;
- maintain discipline;
- know their employees;
- make safety part of everyday business.

Whilst the term may be relatively new, the concept of leadership being *felt* is not. Indeed, looking at the list above we can argue that being able to express leadership so that it is felt by our followers has been one of the attributes of most, if not all, of the great leaders through the ages, whether in safety or more broader business activities.

In the earlier chapter in this book on Leadership, we discussed how great leaders distinguish themselves in some way. Great safety leaders distinguish themselves through their leadership being felt by those around them in a deeper, cognitive way. So, by exploring this concept now with regard to safety, have we found the silver bullet? Well, not quite – workplace safety and risk management will always require more than one tool, procedure, policy or technique – but it is a valid and, for the majority of leaders, a highly potent approach. Felt leadership is not necessarily a new style or theory of leadership, instead we might consider it the distillation of the most impactful elements of leadership. Felt leadership moves leaders from a process of compliance, enforcement and 'telling' to an approach based on shared commitment, by influencing, engaging, empowering and enabling those around them. Over time, many organisations have found that their safety cultures have been considerably enhanced through the application of felt leadership, benefiting not just workplace safety performance, but delivering synergistic value to other areas of the business too.

If we are to attain our goals of safety excellence, and build strong, supportive workplace cultures we need to go beyond speeches on strategy and policy. We must be able to describe in clear and vivid detail what we're striving for in terms of safety and then demonstrate this commitment clearly in our actions. Not enough for a leader to express their commitment to safety in word alone, it is the repeated, consistent action in demonstrating genuine interest in safety and concern for their wellbeing that sets expectation, reinforces behaviour and allows everyone around us to feel our leadership in safety. Visions and missions are useful tools to set direction, but they will only be successful if stakeholders understand what they are there for, believe in them and have the opportunity to support their achievement. If the leader does not lead in safety, how can they expect the workforce to follow safety?

CASE STUDY – ACTIONS SPEAK LOUDER THAN WORDS

The President of an organization I was consulting to was a former military commander, and his leadership style had evolved into a concise, unambiguous, strong style. His sharp focus on driving productivity across the business had ensured that profits were up and shareholders satisfied. Aware that accidents did occur in his production facilities from time to time, the President considered the safety performance to be in line with industry average – after all, they operated in a high-risk sector – and so long as the financial targets were achieved things appeared normal. The President was confident that his commitment to his workforce was evident – he had himself signed the *'Safety First!'* policy statement that hung on the walls of every factory.

When his son was involved in a ski accident, the President's perspective changed. Noticing how much his own life had altered he began to see workplace accidents in a new light. When the next serious injury occurred, at a site well known for poor safety performance, the President cancelled his plans and went immediately to the site to speak with management, the local work teams and the injured party to understand how the accident happened.

During the discussions, he learned that there were several issues with the equipment which encouraged employees to 'take chances' with their own safety in order to maintain the productivity figures. The President carefully listened to what needed to be done and gave his assurance that action would be taken. Within a week the changes were made and the President visited the site again to thank the employees for their suggestions and confirm that sufficient action had been taken. Sharing the story of his son's accident, he explained that he wished to avoid further workplace injuries and invited all employees to raise future safety concerns immediately with their managers and to include him in their requests for support with safety.

On his return to the office, the President instigated a Safety Review Call to be held within 24 hours of any serious injury. The entire leadership team of the business was expected to attend as a matter of priority – and would be held to account for their contribution – and together with the management team from the site where the injury occurred the team, led by the President, they would collectively seek to understand the causes of the event and together identify how to prevent recurrence. Immediately after the call concluded, the President would send a summary of the agreed learning points to the leadership team and all site managers with the express instruction to act at local level.

Within six months a marked reduction in the organisation's accident rate could be observed as safety truly started to come 'first' for the business.

In order for our leadership in safety to be 'felt' by those around us we must hold an absolute personal commitment to a set of shared organisational values on safety. Within your organisation you may already have agreed a set of values, or perhaps through this book you are considering developing some that help guide your approach. Whilst values are certainly very personal to each organisation, frequently there are three specific value statements that are common for many organisations committed to safety excellence. These can provide a useful starting point for felt leadership and are often reflected in the safety principles articulated by leaders:

- All workplace injuries and ill-health are preventable.
- Everyone has the right to go home safely at the end of every day.
- Safety is a line responsibility that is owned, led and driven by the entire team of leaders within the organisation.

In the case study above, the President learns that he needs to change his perspective and *get involved*. Felt leadership is about leaders interacting with people in the field on matters of safety and risk management, collaboratively working together to identify workplace hazards in a positive manner, discussing risks and demonstrating their own personal commitment with questions including:

- 'What can I do to help you make this safer?'
- 'What's the worst thing that could happen here?'
- 'How can we prevent that from occurring?'

Felt leadership is an action-focused, transformational approach to leadership that takes effect through the inter-relationships between leader, followers and those around them. Where practiced systematically, it can have a profoundly positive effect on an organisation's safety culture. The relationship-oriented approach of inspiring trust through the active and consistent demonstration of personal commitment to safety encourages others – no matter where they sit in the organisational hierarchy – to consider their own values and lead from their own hearts. As illustrated by the President in our case study above, the starting point is always the personal commitment to safety excellence that the leader espouses which encourages trust and buy-in from others to adopt these same values and beliefs. This softer, more intimate method moves us usefully away from the directive approach of compliance-driven 'telling' styles of leadership and through our actions our commitment is found to be so strong and ring so clear that it is not just heard or observed, but 'felt' by everyone around us.

Beyond our solid commitment, the second step is to understand the needs of those around us. Research from the field of industrial psychology tells us that engaged employees expect six things from their leaders to ensure that their basic needs are met. These help us to direct our leadership more clearly and are:

1 **Focus me** – set the direction on what's important around here.
2 **Equip me** – provide the skills, tools and resources I need to do my job well.
3 **Know me** – understand who I am and what I have to offer.
4 **Help me see my value** – enable me to understand how what I do is valuable.
5 **Care about me** – don't place me in danger, alert me to risks.
6 **Help me grow** – provide opportunities to stretch and develop.

Effective felt leadership addresses each of these six areas through the interactions that occur between leaders and those around them. When needs are met and leadership is truly felt rather than just heard, relationships take on a more trusting, interdependent, reciprocal style; becoming multi-dimensional and uni-directional rather than the traditional leader-subordinate one-way conversations of cause and effect. The impact of authentic felt leadership encourages a sense of collective responsibility, which over time becomes self-perpetuating and self-governing. The shared responsibility for the attainment of organisational objectives naturally becomes part of everyone's role as leadership for safety becomes distributed throughout the entire workforce.

Let's look now at the actions we can take to demonstrate our commitment to safety.

Leading with safety

We're clear about our values, we've set out our principles for safety, we're committed, and we understand what our people need. What's next? *Action*! Distributed Safety Leadership requires a catalyst for action. This catalyst is *you*!

Great safety leadership that is felt by everyone operates through 3 interconnected pillars:

1 Active leadership
2 Worker involvement *and*
3 Assessment and review.

Within each pillar is a range of organisational, team and safety-related factors that influence actions. Each pillar is as important as the next, and working proactively across all three will ensure not just clarity of leadership, but also go a long way towards attaining your organisational goals and aspirations as well as encouraging others to take their role in the distribution of safety leadership. Each pillar may be worked on independently – however, it's also possible to work across the three concurrently, as we'll see in the case study towards the end of this section.

The pillars of leadership

Each of the three pillars in the model can be broken down into discrete activities, let's look at these in turn now.

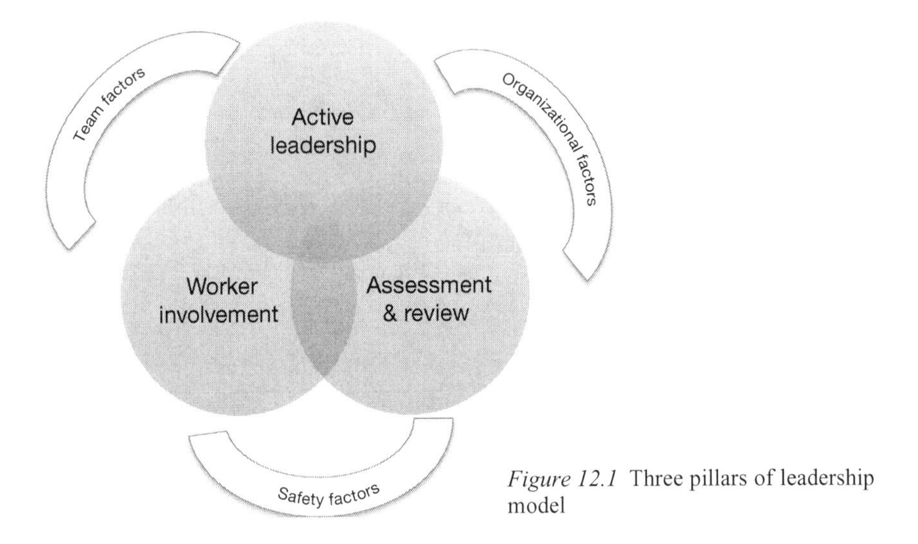

Figure 12.1 Three pillars of leadership model

Pillar 1: active leadership

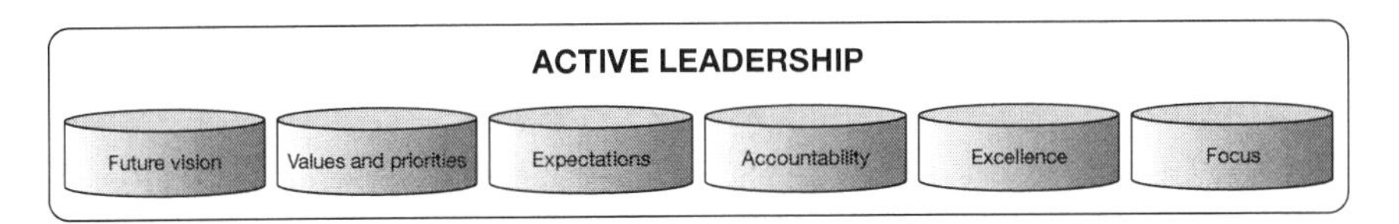

Figure 12.2 Active leadership pillar: cross-sectional slices

Visible, active leadership – from top leadership down throughout the entire leadership and management hierarchies – is imperative. We're not talking about the executives and managers sitting quietly waiting to receive the latest monthly accident figures – but, instead, getting up and out there, into the workplace, and leading the charge, with each tier demonstrating to the one below how to lead with safety. Key activities for the safety leader here include:

Envisioning the future

> "Leadership is about going somewhere. If not in service of a shared vision, your leadership efforts are in danger of becoming self-serving"
>
> Blanchard and Stoner (2011)

Creating and articulating the vision for safety is of paramount importance. No one follows the leader to an uncertain destination, no matter how persuasive they may seem, we have an inbuilt need to know where we're going. The leader must:

- be clear about what the future looks like when the vision is achieved;
- have a realistic understanding of the current organisational approach to safety;
- be able to identify the strengths, weaknesses, opportunities and threats of the current approach;
- be able to see and paint the big picture for safety;
- identify and explain how safety success supports and fuels broader organisational success.

Establishing values and setting priorities

In recent years, many organisations have undergone a sea-change in their perspective on safety, shifting it from being a 'priority' to a 'value'. This is good logic and firmly places safety at the core of the organisation. Beyond articulation of safety as a value, the leader must also be prepared to dig deeper into the safety approach to identify and set priorities for action. Including and integrating good health, safety and risk management with business strategy and decision-making is not just good practice but key to working out which actions need to be taken in order to drive sustainable safety improvement.

Setting expectations

Beyond the articulation of the importance of safety at work, through their words and deeds leaders also set the level of performance they expect. The old axiom "when leaders lead, followers follow" is truly relevant here: consistent behaviour

LEADING BY EXAMPLE

As popular research points out, more than 90 per cent of all workplace accidents are related to some aspect of behaviour. Leaders have a significant impact on the behaviour of workers, both in word and deed.

The old adage that: "actions speak louder than words" rang true for one leader. During a visit to a Scottish shipyard, former UK Prime Minister Gordon Brown declined the invitation to wear a hard hat for the site tour. Imagine the reactions amongst the workforce during the visit, and the following day, when a photograph of the premier together with the organisation's leadership team (all wearing their hard hats) appeared in the national media.

Whose leadership example had the strongest influence? The actions of the PM, or of the site leaders, who naively accepted his declination and permitted a tour of the work area without the required Personal Protective Equipment anyway?

role-modelling coupled with clarity on the standards expected is key to building safety leadership credibility.

Enforcing accountability

If leaders have clearly communicated their expectations, they will expect followers to meet them. Where expectations are not met, appropriate consequences need to be invoked to ensure that procedural justice is delivered. The perennial debate of whether blame should be apportioned in health and safety failings will continue to run its course – however, the *Naked Safety* approach requires that safety is viewed on a parallel with all other business processes. Any smart organisation would hold their people accountable for significant failings in other procedural areas, and safety must be no exception to this rule. Much discussion has occurred in recent years over the notion of 'blame cultures' but to be clear, accountability is not blame. Building a 'just' culture, where everyone is empowered to act, and is held accountable for their contribution is the foundation of great workplace safety.

It is *always* the responsibility of *leadership* to enforce accountability. A normal enough idea, however, when safety leadership is delegated down the line, such as in the 'supportive safety leadership' context described earlier in this chapter, all too often the ability to effectively enforce accountability is not transferred at the same time.

Striving for excellence

Setting expectation and enforcing accountability are both squarely rooted in the command and control style of leadership. Whether expectations are numerical or behavioural, the target is set, and attempts are made to hit the bullseye. But, over time, as people understand how to hit the target, the game becomes one of minimum standards – doing what needs to be done to achieve what's been asked. It's here that great safety leaders move beyond compliance to inspire and encourage others to go above just hitting the prescribed targets and instead push to achieve the best possible outcomes. This notion of discretionary effort is something only the very best leaders can cultivate. It is gained by the leader who leads, and who challenges the *status quo*, not by the other who has been delegated and simply holds support.

Maintaining focus

Through her actions, the safety leader builds rituals that are observed and participated in by her followers. Safety walks around the shop-floor; 'safety moments' to begin each meeting; and 'safety bulletins' to share learning from incidents can all be excellent tools to help maintain a positive focus. But with all repeated actions comes the risk of automaticity: over-standardising a process so that it seems to operate of its own volition and without input. Avoid devaluing your safety leadership activities by keeping things fresh. In his insightful book *The Way We're Working Isn't Working* (2010), Tony Schwartz cautions: "Self-awareness and self-reflection are the antidotes to self-deception." Taking the time to consider and build new rituals can transform our behaviour – and that of our followers.

Pillar 2: worker involvement

Okay, now let's turn to the second pillar, Worker Involvement. Here there are three key aspects to consider.

Figure 12.3 Worker involvement pillar: cross-sectional slices

Influencing and engaging

One of the biggest challenges in safety leadership is mobilising people to address safety and risk challenges. In a recent study DuPont asked their clients: "What's the strongest motivator for engaging employees in safety?" The response was astounding: **Leadership role–modelling** was unanimously agreed upon by hundreds of international CEOs and Presidents as the most effective motivator. A **sense of team spirit** came a close second, and the **opportunity to collaborate** (active participation on OSH activities that brings management and workers together with a unifying objective) arrived in third place. Interestingly, the same business leaders felt that monetary inducements ('carrots') and compliance mechanisms ('sticks') were the least effective.

Effective communication

Great safety leadership revolves around clear, transparent, honest communications on the organisational vision, mission and strategy for safety and risk management, repeated frequently and sincerely to all employees to create and strengthen the corporate cultural glue.

In order to initiate authentic dialogue with others on matters of safety, we need to remember that communication is not just words we use, but also our behaviour, the expressions we employ, our tone of voice, and can even be as subtle as how we make eye contact with our audience. Great leaders are aware of their non-verbal communication as much as they are the words that they say. Effective communicators:

- communicate clearly and with urgency on safety matters;
- listen actively to the contributions of others;
- ask questions to engage the audience and ensure dialogue rather than monologue;
- focus on the positive, turning problems into challenges to be solved collectively;
- adopt a caring, humanistic approach to discussing safety issues;
- use corporate values to connect safety to the values of individuals, groups and teams;
- advance safety communications beyond their immediate sphere of responsibility by reaching into the supply chain, contractor groups and wider stakeholders;
- always have ready a clear and concise story, statement or conversation-starter on safety.

'MANAGEMENT BY WALKING AROUND'

Since its introduction as a concept in the early 1980s, 'Management by Walking Around' has enjoyed a rollercoaster ride in terms of perceptions on its utility, its application and its effectiveness. But in times of change or uncertainty, when conducted with an authentic and sincere interest in the opinions and contributions of others it does provide an excellent opportunity to engage others, offer encouragement, build confidence and inspire action.

Each time the leader ventures onto the shop-floor is an opportunity to engage with individuals or groups of employees and use their influence to reinforce the organisation's commitment to safety as part of its values system. By taking these opportunities, the leader clearly demonstrates what's important and what is expected of them as individual contributors. Naturally, where the most senior leader is conducting such activity, this quickly sets the tone however when the whole leadership team carry out these visits, results are not just multiplied, but accelerated.

Communication – the point of safety

As organisational leaders assume responsibility for safety, they begin to think about how they can personally demonstrate their commitment and show leadership of this worthy cause. Oftentimes, this may manifest with activities such as beginning each meeting with a safety point for discussion or to focus the minds of attendees on safety. Though this is an excellent start if not already in place, all too often, leaders 'go through the motions' with this exercise and within a short period of time it becomes a rather hollow activity where little value or benefit is derived. But the leader has become hamstrung by their own best intentions – stop the concept now and it looks like his commitment has waned. So, what to do?

REFLECTION POINT – ELEVATOR PITCH

Imagine you've just stepped into the elevator and you find the CEO on the way to an important meeting. *(If you don't have an elevator in your office, you may need an extra bit of imagination here!)*

The CEO turns to you and explains that he needs to give an update on safety and asks you for your thoughts. As you watch the elevator lights flash closer to the CEO's floor what do you say?

Why do you do safety in your business?

How does the workforce know that safety is important around here?

What value does safety bring to your organisation?

The elevator stops and the CEO thanks you for your ideas. As he steps out he mentions that the meeting will be attended exclusively by 300 shop-floor employees. The elevator doors close and you begin to descend to your floor. Did you give the CEO the right steer? Would your 'elevator pitch' be different had you known who the intended audience was?

What's the (safety) point?

'Safety Points' (also commonly known as 'safety contact', 'safety flash', 'safety minute', 'safety moment' or a range of other terms) are short information sharing sessions typically taking just a few minutes at the beginning of a team meeting, briefing or other get-together. By beginning every event with a focus on safety, the organisation (and the individual leader) demonstrates a commitment to safety being part of the way business is done, and highlight how importantly safety is considered by top management.

Leading the Safety Point can be a daunting task – many organisations charge leaders with the duty, but allow complete free reign as to how the leader constructs and implements the Safety Point. Without guidance or content, delivery can be variable, and over time, the value of these moments erodes until a group realisation kicks in and the activity slides off the agenda.

Great organisations see the true value of these sessions and take care to provide topics for delivery, and some go as far as setting a programme of topics for the year ahead. However, in order deliver a high-value, impactful safety point such structure is not *essential*; a powerful point can be delivered using the following five-point framework.

- **Topic**: Select a topic that is relevant to the task, activity or work programme being done.
- **Content**: Safety Points do not have to originate inside the workplace. In fact, many of the best safety points I've heard have come from outside the organisation. When you're reading your newspaper over breakfast, or listening to the radio on the drive to work or perhaps even chatting to friends, what snippets of information catch your attention? Would a news story about injuries to cyclists help you with a safety point about road safety for your people driving delivery vehicles? Can the opening of a new fitness centre locally give you a reason to talk about the importance of physical fitness or warming up for manual labour tasks? Could a radio bulletin advising a spillage from a chemicals truck give you a route into talking about how your team would manage a spill at work?
- **Format**: Many organisations prefer to use a small set of slides, or a paper-based copy, but this can encourage the leader to *over*-develop the point. Remember that the purpose of the safety point is to focus attention and get people thinking for themselves – it's not a training exercise so keep it light, keep it brief.
- **Delivery**: Simple story-telling styles are easier for most people to comprehend and remember. Keep the content concise to allow time for debate and discussion on how the safety point content may apply in the workplace. Identify how discussion of the safety point supports or links to your corporate values.
- **Questions**: Use questions to underline the key points and to help the team understand what can be taken from the story and used in your workplace. Ask questions to the team, and encourage them to ask questions amongst themselves. You may find that the team agree a particular focus, or action to take forward themselves.

Teamwork

Safety is a team game. But for some people it may not be the most appealing game to play, so how do we encourage people to work together in the promotion and achievement of safe working conditions? A solid first step is to ensure that appropriate training, instruction and supervision is in place to allow employees to understand the risks associated with their jobs, and how to manage these safely.

Daniel Goleman suggests that, in general, those people who are currently working together in a team will already be attuned to each other and operate in congruence, anticipating the actions and reactions of those around them.

The most effective teams are the ones that have high levels of 'team harmony'. Goleman identifies the six norms that maintain harmony:

- Team members are aware of each other strengths and weaknesses.
- Teams let their members step into or out of specific roles as needed.
- They don't let friction keep simmering away until things start to boil over.
- Dissent or dissatisfaction is dealt with it before it becomes a problem.
- They have a great time working with each other.
- Harmonious teams celebrate wins – even small wins – together.

An old friend of mine, Jim Weigand – former President of DuPont Sustainable Solutions, is a big fan of teamwork and believes that effective, impactful teamwork goes beyond the act of cooperation. Jim's mantra is that good safety is all about collaboration. Ask him to do some mental arithmetic and Jim will explain that in his 'Collaboration Equation' two plus two always equals five. When people collaborate, the result is greater than the sum of the parts.

THE COLLABORATION EQUATION

2 + 2 = **5**

So, how do we obtain the higher impact result that collaboration offers? Weigand's view is remarkably straightforward, with five pre-requisites to facilitate collaborative working:

1. Set clear objectives for the team.
2. Encourage peer-to-peer relationships.
3. Give ownership to the team.
4. Allow sufficient time for the team members to build trust.
5. Act with urgency to master conflict as it arises.

Pillar 3: assessment and review

And finally, we turn to the third pillar to consider the activities of Assessment and Review. Here we have six key elements to discuss.

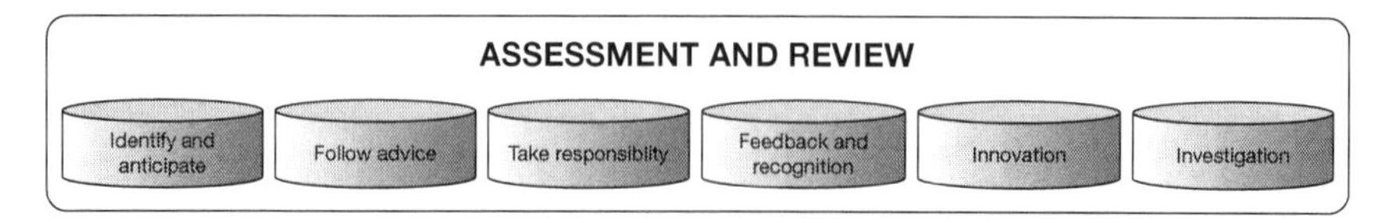

Figure 12.4 Assessment and review pillar: cross-sectional slices

Anticipate, identify and assess workplace H&S

As we examined the key elements of felt leadership at the beginning of this section we learned that it's crucial for leaders to 'know the operation'. A solid understanding of how things work is a great foundation for thinking about safety risks. This doesn't mean that every leader needs to become a technical safety expert! Often, a fresh pair of eyes can identify risks where they may not have been noticed by those working in the area. A genuinely inquisitive approach accompanied by an honest question asked in the spirit of learning can be highly effective in opening a dialogue on safety, and lead to dynamic real-time risk assessment.

Accessing and following competent OSH advice

Great safety leaders know and accept that they are not the experts in the field. They understand that their purpose is to lead, to engage, to inspire action. These leaders recognise the importance and value of competent advice, and regularly seek the input and guidance from technical experts to ensure that they remain on track and aligned to the aims of the safety strategy.

Responsibility for performance

The traditional marital vow: "for richer for poorer, in sickness and in health" rings true for safety leadership too. The great safety leader is the one who keeps her eye on what's going on, reviewing safety performance and progress regularly and with genuine interest. When the storm clouds roll in, she steps up to accept responsibility and takes the lead in working out what needs to be done to get back on track.

Feedback and recognition

Cultures may dictate how feedback and recognition is applied and received but the great safety leader remains mindful and sensitive. In a room filled with several hundred shop-floor workers, I observed a leader spend a full five minutes eloquently describing the wonderful actions he had witnessed whilst touring the factory, speaking so highly of the personal commitment he had witnessed in a machine operator.

After his fine words of deeply expressed gratitude, the leader named the individual and asked them to stand up and come and join him on the stage. Already feeling more than a little embarrassed for such overwhelming praise for what she had thought was just 'doing her job,' the employee was reluctant to leave her seat amongst her colleagues. Despite the repeated invitation to join him on the stage, the leader remained alone, and the employee sat tight. Stuck in the spot between embarrassment and ridicule the employee later admitted that she had felt helpless, and detested the idea of such glorification in front of her co-workers.

FEEDBACK

Ken Blanchard believes that "Feedback is the breakfast of champions." Whether Olympic athletes or business leaders we can all recognise the importance of fuelling up well before we start our day. But how many times do we skip this important opportunity? One of the most important tools when it comes to developing distributed leadership in safety is feedback. Done well, it strengthens the relationship between two people. But it's also one of the easiest things to get wrong. What starts with good intentions can quickly plummet to being perceived as criticism and lead to feelings of inadequacy and resentment, effectively demotivating rather than encouraging the recipient. Getting the balance just right is crucial, so take a step back, think carefully about the action that you want to inspire, and then consider the words you'll use to convey the message. Whether providing positive or negative news, remain sincere and don't fall into the trap of apologising for what you are saying. Remember that feedback is a gift, so take care to choose the content carefully to match the recipient, and present it in a clean and tidy fashion.

RECOGNITION

Recognition can be a route to rapid engagement of employees with the core values of the business. Research time and time again shows that employees who feel appreciated for their efforts are amongst the happiest, and most enthusiastic supporters of the organisation in its pursuit of its core values, and will go that extra mile in terms of discretionary effort. Conversely, those employees who do not feel recognised or feel under-valued are found to hold negative attitudes towards the company. It is estimated that up to 50 per cent of an organisation's workforce can feel under-appreciated.

Recognition schemes can provide practical and inexpensive route to encouraging commitment and buy in from the workforce towards organisational values and share the vision for the future. When linked to the core values of the organisation they can help to drive the focus and provide direction to the entire workforce.

It's important to remember that recognition does always not have to be a public display of gratitude, nor result in gift or reward-giving. Recognition at its most stripped back is about being aware and appreciating the resourcefulness, innovation or persistence it takes to get a job completed safely. When considering how you as a leader recognise contribution, bear in mind that for most people true recognition can be a simple as:

- to be shown respect through verbal and non-verbal communication and behaviour;
- to be recognised for a personal contribution to wider operation of the organisation;
- to receive the opportunity for understanding, support and two-way dialogue.

CASE STUDY: RECOGNITION IN PRACTICE

An engineering firm implemented a *Rapid Recognition* scheme to reward employees who offered a good idea to improve safety or whose actions made an important contribution. Prizes varied from baseball caps featuring the company safety logo, to tickets for sports games, dinner for two and vouchers for local stores. Each month a special lunch was help in the works canteen, hosted by the senior leadership team, to recognise star performers who had made an extra-special contribution towards the company's core values.

A global manufacturing organisation rewarded all employees at each site that had achieved a full calendar month worked without an injury to an employee. Free breakfasts were awarded to everyone on site. Many sites enjoyed free breakfasts for several successive months. When a site achieved a full year without injury they received an open letter to all employees from the CEO, thanking them for their commitment to safety and inviting them to celebrate with a party at the site. Accompanying the letter was a large personalised banner reflecting the achievement, signed by the CEO, which the site could display prominently.

Formal award ceremonies can be excellent opportunities to reflect corporate core values, and when safety is mixed in with other 'essential' aspects of the organisation, top safety contributors are rightfully celebrated right across the organisation and gain broad visibility amongst their peers. One organisation operated an annual awards scheme covering five disciplines (Safety, Quality, Productivity, Customer Service, Performance Improvement). After three years of running the awards, the organisation found that submissions to the Safety category outweighed submissions to the other four categories by seven to one!

Innovation

NEW IDEAS

Some advice I received early in my career in safety was that there are 'no new ideas in safety – everything has been done before'. But despite this assertion, in my own career in safety and risk management I've been fortunate to observe many great new ideas and perspectives. In fact, one of the world's most prestigious and longest running international safety awards programmes believes so strongly in the importance of innovating in safety that they created a specific category of their awards programme to recognise this very topic. Interestingly, over the last decade, the volume of submissions to the 'Innovation' category has continued to rise, year on year, underlining that there are indeed many new ideas in safety.

RECYCLING IDEAS

In recent years, we've been encouraged to recycle as much as we can, both in our private lives and in the workplace. Look around yourself right now and chances are you'll spot recycling in action. Paper, coffee cups and much more. The maxim 'don't reinvent the wheel' is a good one, and extremely relevant in all aspects of business life, not just in safety and risk management, and we can find that many ideas we have may already be in place elsewhere – learning from others' successes and failures is an important part of life. Steve Jobs, former co-founder and CEO of Apple Computer famously remarked:

> Picasso had a saying: "Good artists copy, great artists steal." We have always been shameless about stealing great ideas . . . I think part of what made the Macintosh great was that the people working on it were musicians, poets, artists, zoologists and historians who also happened to be the best computer scientists in the world.

In their revolutionising of the home computing and personal mobile telephone industries, Apple made no secret of the impact of cross-functional collaborative working had on their products design and utility. Sometimes, the best ideas can come from where we least expect. The next time you're trying to come up with a new safety idea, where will you find your 'zoologists and historians'?

Investigate incidents

When things go wrong, and they will at some point, even in a perfect world, the committed safety leader gets involved. Don't leave the investigation to the line management, or the technical expert, get amongst the action and be prepared to ask questions. When the Safety Review Calls began in the organisation

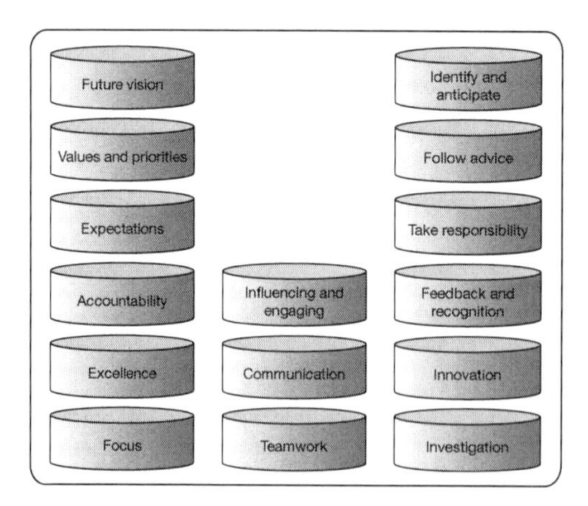

Figure 12.5 Three pillars of leadership model: cross-sectional slices

mentioned in the case study above (*'Actions Speak Louder Than Words'*) many of the operational leaders remained silent. When asked why, they revealed that they didn't feel as if they knew the technical aspects of what had occurred so felt it best to keep quiet. Just like a fresh pair of eyes can be beneficial in workplace inspections, fresh thoughts, questions and ideas can all add sizeable value during investigations. Remember that great safety leaders check their ego at the door and don't worry about risking their reputation when it comes to leading from the heart on safety. Ask the naïve question. Offer the simple idea. Suggest a corrective action. If they are rejected, it's no great loss. If they're never mentioned and you have the same thought again when a similar incident occurs, the personal regret will be significant.

By working through each of the three pillars of felt leadership, we can understand the essential elements of leading with safety. Practiced regularly, these actions clearly demonstrate your personal commitment and help to strip back some of that mystery that can often surround safety.

There are no rules for leading with safety: you may choose to engage with each activity one at a time, in a stepwise manner. Alternatively, you might elect to consolidate a number of the elements at once, as the organisation in our next case study did.

CASE STUDY – WALKING THE TALK: THE THREE PILLARS OF FELT LEADERSHIP IN PRACTICE

A global engineering company identified that their safety performance was not where they desired it to be. To commence a step change in their approach, the CEO invited the workforce to a 'town hall' meeting where she set out her views on why safety was important to the business, how the policy and strategy had been revised and what her expectations were.

Anticipating the introduction of a new safety initiative or campaign, the workers were surprised to hear that this new approach would be driven by the senior leadership team. The CEO explained that each month the executives would separately go to production facilities across the organisation to conduct 'Safety Leadership Visits,' which would involve the leaders visiting the shop-floor and talking with workers to learn more about safety to understand how they could help to improve things at local level. The executive team (of 12 direct reports) would each be charged with conducting one visit per month, and to underline her own personal commitment, the CEO set herself the same target.

At each monthly board meeting, the CEO would review progress and feedback to the leadership team, and then share a summary of the visits with the business.

Over the next few months, imagine the response from the production sites when the Corporate Legal Counsel, Marketing VP or Director of Finance arrived in the machine hall or workshops, dressed properly in their protective equipment, and began to discuss safety issues directly with the workforce.

Very quickly, it became clear right across the organisation that there was a renewed focus on safety, this time, reinforced by real and tangible commitment from the entire leadership team. After 12 months of running the Safety Leadership Visits, qualitative feedback obtained via the annual employee opinion survey and quantitative data on accident rates confirmed that a radical shift had occurred within the culture of the organisation. Encouraged by the success of this new approach, a formal Safety Leadership Visit is now part of the regular routine for all leaders, managers and supervisors.

Safety leadership visits – a practical application of felt leadership

The process for conducting safety leadership visits can be very straightforward (if you prefer to make things really complicated, go ahead, though the impact may not be as positive), after all, the entire ethos of *Naked Safety* is about stripping things back and making it as simple as possible.

Prior to entering the work environment or the shop-floor, the leader takes time to ensure she is wearing all the necessary Personal Protective Equipment and complying with any local rules (for example to remove jewellery, tie up long hair, etc.). Before approaching anyone, she observes what's going on, looking out for any immediate hazards – slippery floors, moving vehicles or machinery, etc.

1 **Introduction** – The leader identifies an appropriate moment to approach an employee (or small group) and introduces herself with a statement like: "Hello, I'm Jane Smith and I'm the Vice President of Marketing here. I can see that you're currently making X widgets and I'm interested in how you do that safely. Can you tell me a bit about that please?"
2 **Listen** – American crime writer Margaret Millar offered an outstanding perspective on the value of listening when she said: "Most conversations are simply monologues delivered in the presence of a witness." The leaders in our case study above were mindful of this, and took care to listen actively to the explanations given by employees, often reflecting or 'mirroring' what they had heard.
3 **Ask** – Using the information that had been provided, the leader then asks further questions related to the task in hand, for example: "If I were to help you here today with this job, what would I need to know in order to be safe?"; "What controls are in place to ensure that these moving parts / hazardous

chemicals / fragile components*(delete / insert as appropriate) don't cause us harm?" Other questions that link activity to the safety vision are also included at this stage, including: "How do you think that working safely helps our business?"; "Can you tell me why safety is important to you personally?"

4 **Expand** – To broaden out the conversation from the current task to a more holistic view, the leader might ask about safety performance, using questions like: "What's the current accident rate for this line/department/site?"; "How do you think that compares to the rest of our business/industry?"; or "When was the last accident around here?"; "Can you tell me what happened?"; "What were the learnings gained from that event?"; "How do these learnings help us to work more safely now?"; or perhaps around culture: "How do you talk about safety in this department?"; "When was the last time your team held a toolbox talk on safety?"; "How does your team leader manage safety around here?"

5 **Review and reinforce** – Following the discussion, the leader recaps what she has learned about the risks of the task, working safely and other key points shared. Before closing, the leader recaps the organisation's commitment to safety by explaining why safety is important – to them personally, and to the business – and reminding the individual of the vision for safety. The ultimate question reinforces her personal commitment to safety: "What can I do to help you improve safety even more around here?" This final request may require some gentle digging, but in our case study organisations, leaders found that employees quickly had suggestions in more than 90 per cent of visits. Thanking the employee for their help the leader safely leaves the work area.

In our case study, the organisational leaders were asked to conduct one safety leadership visit each month. A one-page visit template was created by the safety manager to provide an outline of the actions and included a range of suggested questions that could be used. The template was never shown during the visit, but was often referred to before the dialogue. Following the visit, the leader added a few lines of comment to the template and shared this with the site manager for action follow up. Some of the most effective leaders involved in this process carefully kept their own notes of who they spoke to and during subsequent visits back to that site, would seek out the individuals to thank them and let them know how they had followed up on their suggestions and check that the anticipated improvement in safety had been achieved. During these informal follow-ups, it became common for other employees to join the impromptu discussions. The critical mass was forming, and 'distributed safety leadership' was taking hold.

The value of consistency

Leading with safety isn't a 'black art': it's good business leadership. The felt leadership approach is an important aspect of developing a positive organisational culture that values the safety of its people. Employees pay most attention to what they observe their leaders doing systematically so demonstrating our personal commitment by *systematically* engaging others in dialogue on safety matters, investigating accidents, conducting safety reviews or risk assessments, is key. Regular, consistent actions with a specific focus stand out in the eyes of those around us, clearly articulating through our behaviour just how important the safety and wellbeing of our people is to us. This builds confidence and inspires action, and crucially, sets the ball rolling regarding the concept of a distributed leadership in safety.

Felt leadership is fundamentally about being true to our commitment. Living up to our word. Felt leadership is the embodiment of our safety vision and its enactment through our behaviours. The Three Pillar Model shines a light on the interplay between taking action, involving those around us, and being reflective.

THE NAKED TRUTH: LEADING WITH SAFETY

Every leader has an influence on safety, whether they intend to or not. Leadership is, after all, everything we do or don't do, and everything we say or don't say. Some leaders will impact more directly the day-to-day performance and activities; others will have a greater influence over the wider organisational culture. No matter in which area we work, as leaders we must recognise that we have a responsibility towards building a culture that recognises the importance and value of safety.

Effective safety leadership requires the courage of conviction to stand up and take action. Stand out, be different, be *felt*. Your role is to be the catalyst for action and show what needs to be done.

The most effective safety leaders don't take an interest in workplace safety because the law requires it, or even because it's the 'right thing to do' for their people. These leaders focus in on safety because they understand and observe that enhanced safety catalyses improvements right across their domain. By treating safety with their *absolute commitment*, conveying a *strong sense of urgency* and acting with *discipline* these leaders not only demonstrate their beliefs and values, but they *inspire* others to join them. It is these three characteristics that build the critical mass that encourages a culture of safety to flourish and ignite the intrinsic motivation of those around them. Imagine, in your own organisation, what it would be like if each of your leaders worked in this way.

Great safety leaders, no matter where they stand in the organisational hierarchy, differentiate themselves through a distinct mindset. These leaders think like winners and have an attitude that is clear to all: they lead from the heart and are determined to succeed. No matter their discipline or area of responsibility, these people manage their task or work function with a clearly visible commitment to the safety of every employee, responding to issues, concerns and opportunities in the workplace with an unequivocal, relentless commitment to managing risks and doing things safely. For these leaders, there is no room for compromise; everyone working safely is their way of life.

Distributing safety leadership across all areas of the organisation, at every level, can rapidly develop an innovative, top-down, bottom-up, side-to-side culture where everyone is engaged, fully focused and committed to sustaining a culture where safety excellence is an attainable goal.

Leaders who truly believe in safety are never starved of meaning in their work; they *'face everything'* and *'avoid nothing'* as they authentically demonstrate their belief that *everyone* should return home safely at the end of every working day. This is not just heard, but felt by everyone around them.

NAKED CHECKLIST: LEADING WITH SAFETY

1. How do you demonstrate the organisation's commitment to workplace health and safety through your personal leadership? How is your personal safety leadership *felt* in your workplace?
2. Which values do you use to guide you in your daily leadership decisions and actions? Are your personal values aligned to the safety behaviours you wish to see in those around you? What actions do you need to take now to bring your safety values to life?
3. As a leader, do you understand the main risks inherent in your workplace and work activities? Are you able to articulate these in your conversations with workers?
4. How confident are you that there are sufficient processes in place to consult with and provide forward communication on safety to your workforce? Where are the opportunities to improve communication?
5. What information does the board receive regarding health and safety? What impact does this have at board meetings? Does it drive discussion and debate? How is it used to make decisions and plan actions?
6. Steve Jobs said: "Good artists copy, great artists steal" in reference to his focus on always looking beyond his industry for new ideas. Where can you look – outside of your sector, and beyond the world of safety – for your next big idea?

Notes

1. Institute of Directors & Health and Safety Commission (2007) Leading Health and Safety at Work: Leadership Actions for Directors and Board Members. HSE, UK.
2. You'll find more on organizational values in the chapter in this book on Corporate Culture.
3. If you haven't yet identified corporate values for safety, you'll find some suggestions coming up in the next few pages.
4. See the chapter on 'Leadership' in this book.

13 Engagement

The value of worker engagement for organisations cannot be overstated. When delivered effectively, employee engagement strategies can not only boost morale and job satisfaction, but also reduce staff turnover, improve worker safety and aid efficiency rates.

In this chapter, we'll look at how an organisation might successfully implement an effective engagement strategy, that will energise the workforce and instil a greater sense of purpose, vigour and enthusiasm.

Resistance to learning is frequently found to be one of the biggest threats to improvement of workplace safety. Often when organisations look at the data they've captured around safety, if the information they've collated in any way conflicts with the preconceived ideas of where they believe the dangers exist, their first response is simply to challenge the data. This is not uncommon; it's a prevalent issue that spans multiple industries and sectors.

The problem here is essentially two-fold. There is often a lack of understanding with regard to how to correctly interpret the data collected – but, also, regularly a casual disregard for the importance of that data. As a result, in many organisations, safety actions may be based on little more than the 'gut feeling' of a nominated expert.

Instead of relying on the insight of one individual or one team, organisations should proactively search for deeper understanding and smarter solutions. The most powerful way to achieve this is also extremely simple: by fostering an atmosphere where problem-solving is not just encouraged and valued, but, crucially, is something that engages the entire workforce.

Engaged workers are safer workers

Worker engagement is not just an effective barrier to accidents, it's also perhaps one of the greatest tools when it comes to *creating safety in the workplace.* Research shows that engaged workers are far more likely to use their initiative within the workplace and significantly more likely to propose improvements to safety systems (Ronald, 1999). A recent study by workplace management and wellbeing expert, James K. Harter, investigating the link between worker engagement and safety outcomes, revealed that the top 25 per cent of organisations – based on engagement – have 49 per cent fewer safety incidents than the bottom 25 per cent (Harter et al., 2009). Conversely, when workers are not engaged, they are liable to be less focused on their work, and consequently more apt to make mistakes.

In practical terms, full engagement means fostering a sense of community and instilling an atmosphere where every worker, at every level, feels able to actively participate – and free to challenge. Whilst managing workplace conditions and improving processes will be effective, it's only when the entire workforce is engaged that most organisations are able to make any real step forward in safety performance.

When workers feel engaged, it adds ownership to their roles and imbues them with a greater sense of responsibility – and, in the workplace, attitudes always drive performance and behaviours.

The Society for Human Resource Management (SHRM), an organisation that analyses business practices for HR professionals, recently revealed that, in a large manufacturing company they were investigating, workers who were deemed 'engaged' were five times less likely to be involved in 'a safety incident' – and seven times less likely to have 'a safety incident involving time off work'.

Respect

In his excellent book *Carrots and Sticks Don't Work: Build a Culture of Employee Engagement with the Principles of RESPECT* (2010), Dr. Paul Marciano, an authority on employee engagement and retention, illustrates strategies to increase worker engagement and reduce organisational turnover through a number of real-life case studies.

The book also showcases Marciano's acclaimed 'RESPECT Model' – a system made up of seven easy-to-implement elements:

- Recognition
- Empowerment
- Supportive feedback
- Partnering
- Expectations
- Consideration
- Trust.

Many organisations operate systems where recognition and reward are important parts of development – but, according to Marciano's research, it shouldn't stop there. This is because recognition – which credits a way of working – is not concerned with worker engagement.

Whilst it is often the case that workers will change patterns of behaviours in order to achieve interim goals, once the goal has been reached they will naturally shift back into their habitual behaviour patterns.

Recognition and reward systems tend to fall short simply because they usually do little to change the safety culture within an organisation – but the problem is larger than that. The really big problem with reward-based systems is that it contributes to festering resentment amongst workers when they don't receive them. Whereas, in the organisations that routinely hand out rewards to all workers to avoid this, they soon become expected – and are, as a result, rendered completely meaningless.

The recognition and reward system can also be dangerous. Systems that reward good behaviour, will also punish bad behaviour – hence the carrot and the stick. However, the problem of the stick encourages a mindset of not getting caught. Clearly, in terms of organisational safety, for example, this is not a good thing.

Marciano's 'RESPECT Model' identifies seven critical drivers that strongly influence worker engagement. So, let's look at them in greater depth:

- **Recognition:** In highly-engaged working environments, workers will feel their contribution and hard work is duly recognised. Team leaders will make a point of acknowledging outstanding performance and reward workers for going the extra mile.
- **Empowerment:** Engaged workers experience increased levels of autonomy. Leaders provide them with the necessary tools, resources and training – thus,

empowering them to succeed. Greater levels of communication between leaders and workers ensures they continually reach goals.

- **Supportive feedback:** Leaders deliver specific feedback in a in a sincere, constructive and timely manner. This is always used to reinforce an improvement – not embarrass or punish.
- **Partnering**: Workers and leaders meet, discuss and collaborate in order to achieve goals. Interdepartmental bonds are healthy, and communication is strong throughout the organisation. Workers in all departments and at all levels share information with one another.
- **Expectations:** Goals, objectives and business priorities are established, achievable and well-communicated. Throughout the organisation, workers understand the standards by which their performance is evaluated. Workers are accountable for meeting their performance expectations and are aware of this.
- **Consideration:** Regardless of their position within the organisation, workers demonstrate goodwill, empathy and care towards their co-workers. Leaders proactively seek greater understanding of their workers' thoughts, opinions and concerns – and are considerate and supportive when they experience personal issues.
- **Trust:** Leaders demonstratively have trust and confidence in the workers and their abilities. At the same time, workers trust that leaders are proficient in their roles– and the direction they are moving the organisation in is the correct one.

Marciano's model is 'actionable philosophy' which, when properly implemented, can help to create transforming and enduring change at both the individual and organisational level. It's founded on the premise that healthy relationships only happen within the context of a respectful environment. So, what else can we do to create that?

Acknowledgement

Everyone wants to have their contribution acknowledged. Workers in your organisation are no different. They want to feel as though their good ideas and daily toil are appreciated.

As well as being critically important as a motivation tool, acknowledgement is also extremely useful for highlighting and reinforcing the sorts of behaviours that will lead to success.

To be effective, acknowledgement needs to honest and direct. Leaders who are adept at acknowledging the contribution of others will ascertain what is important to the worker and then acknowledge this within the wider context – stressing how it supports the values or goals of the organisation. Here's an example:

> Leader: "I just want to acknowledge you for reporting that near miss. I know it's easier just to let these things lie – not to mention a lot less paperwork – but because you reported it, that data can now be added to a review and we

can look at ways to stop it happening again. At the end of the day, we value safety here – and without accurate reporting of all of these sorts of incidents, we're basically flying blind!"

Though the two are often confused, acknowledgement is very different to praise. For example, praise is often vague and doesn't reinforce specific behaviours or skills. Nor does it explicitly link them to future goals.

Acknowledgment, on the other hand, is specific. It's applied strategically with the intention developing workers and highlighting how their current competencies and behaviours are aligned with the aspirations of the organisation.

Where a lot of leaders naturally let themselves down when it comes to acknowledgement, is that they fail to make it sincere. People are generally quite savvy and very accustomed to reading others. So, if an acknowledgement sounds mechanical or perfunctory, this will almost certainly be picked up.

Good leaders take a genuine interest in the people that work for them – and when they acknowledge a member of the team it will naturally come across as sincere. When leaders try to fake it, the words won't accord with their tone of voice, their body language or energy level and the whole thing seems forced and unnatural.

Acknowledgment builds rapport and fosters trusting relationships. Acknowledging another person with sincerity is difficult without a strong connection. To do so, it's necessary to be cognisant of their skills and abilities and what their goals are. You must also understand the challenges they face on a day-to-day basis and why the behaviours or skills they've demonstrated represent personal growth.

Acknowledgment is a cornerstone of appreciative inquiry—a proven method of positive reinforcement. Sincere acknowledgment reminds workers that what they're doing is appreciated and reinforces how it relates to the organisation's objectives. It also strengthens relations by demonstrating that the leader cares about the worker's contributions and future success. Here are four golden rules of acknowledgement:

1. **Use specific examples:** be clear about the effort being acknowledged. Not only will this demonstrate a leader's personal investment in what has taken place, but it will also encourage other workers by reinforcing that desirable behaviours and contributions are recognised and encouraged.
2. **Don't use superlatives (e.g. 'great' or 'fantastic'):** when leaders get into the habit of overusing blandishments, they soon lose their power and become meaningless.
3. **Acknowledge behaviours or actions that are meaningful to the recipient:** this is a good way for leaders to show that they remember the goals and objectives of workers and, crucially, anything that might be worrying or causing them personal difficulties.
4. **Be sincere:** make a concerted effort to invest time and energy with people and communicate in a natural, steady way that shows that you care. A rushed

comment whilst passing someone will have less positive impact than you taking the time to meet with someone and look them in the eye as you acknowledge their contribution.

Motivation

Motivation is a psychological concept defined as the process that initiates, directs and maintains goal-orientated behaviour.

The definition put forward by American psychologist, Marshall R. Jones, in *Current Theory and Research in Motivation* (1953) remains popular and is still regularly quoted, despite half a century of subsequent research.

Jones described motivation as: "how behaviour gets started, is energised, is sustained, is directed, is stopped – and what kind of subjective reaction is present in the organism while all this is going on".

Motivated workers work more efficiently; the work is completed with greater efficiency, with higher levels of collaboration, creativity and commitment – all of which has a positive impact on the bottom line.

Motivation and engagement

There is an intrinsic link between engagement and motivation – with many modern theorists believing that motivation is merely a function of engagement.

A recently conducted study of businesses in 142 different countries across the world – on *The State of the Global Workplace* – by Gallup.com, suggested that very few workers feel engaged at work. Out of the staggering 180 million workers Gallup polled, only around 13 per cent claimed to be 'psychologically committed to their jobs' enough that they would routinely make a positive contribution to their organisations.

Experienced leaders know that a motivated and engaged workplace performs to best of its abilities. Without motivation, workers lose sight of how their contribution matters. They quickly come to feel underappreciated and disenfranchised, simply because the opportunity to link their passion to the work is lost.

Disengagement also works like a virus – meaning that workers that are not motivated worsen performance prospects simply because they tend to bring everyone else down.

Setting goals

All organisations, no matter what industry or sector, share one single trait – they want to succeed.

Whatever an organisation's function, success is always the primary goal – and, consequently, goal-setting remains a key motivational technique.

Goal setting, aligning and tracking are critical to success – and, when closely tied to the organisation's overall strategy, can see worker performances improve exponentially.

Engagement within organisations is not simply about leaders and managers finding out what their staff are doing at the weekend or sympathising about a sick pet gerbil (though these are good ways to build rapport, obviously), it's more a process of working with them in order to visualise success – and their part in it.

Setting goals should be a collaborative process between worker and manager, and all goals need to be detailed, quantifiable and, yet, easy to understand. A popular method of setting objectives is by utilising the S-M–A-R–T framework.

- **Specific:** Workers need to have complete clarity of what is expected of them. It is a manager's role to create specific, defined goals so that success or failure can be easily measured progress can easy tracked by both worker and manager.
- **Measurable:** Managers need to provide milestones in order to track progress and motivate them towards success.
- **Attainable or Achievable:** Goals needs to be realistic and agreed by both managers and workers.
- **Relevant:** Goals need to be focused so that they have maximum impact to the organisation's overall strategy.
- **Timely:** Managers need to ensure that the deadlines set for goals are realistic but not so long that they undermine performance.

Equity theory

Developed in the early 1960s by behavioural psychologist, John S. Adams, 'equity theory' is about defining and measuring of the 'relational satisfaction' of workers. In terms of business psychology, equity theory comes under the umbrella of 'organisational justice'.

The theory focuses on determining whether the distribution of resources is 'fair' – and is measured by comparing the ratio of contribution and reward for each worker.

Adams believed that workers will naturally strive to maintain equity between the efforts and skills that they personally bring to their job and what they receive back, against the perceived efforts/skills and rewards of those about them.

The central principle behind equity theory is that workers value fair treatment – and the effort to maintain the fair relationships of their co-workers, and the organisation, will keep them motivated.

Equity theory recognises that a number of subtle factors can influence an individual worker's perception of their relationship with their relational partners. However, according to Adams, financial reparation is typically the cause of equity or inequity. Perceived underpayment is often a cause of significant anger and resentment, whereas, overpayment induces worker guilt.

'Equity norm' is what an individual worker expects as the fair return for their contribution; which is determined by comparing their inputs with that of their co-workers – a process Adams refers to as 'social comparison'.

Though it is a key factor, it is not always based on financial issues. For example, if a factory worker is harshly rebuked by the site manager for not reporting a near miss, but knows that a lot of her co-workers have also previously failed to report similar incidents, the perceived unfairness of the situation is likely to be the cause of antipathy.

When a worker perceives – rightly or wrongly – that they are in an inequitable situation, they will generally respond in one of three ways:

- By reducing the inequity either by distorting inputs and/or outcomes in their own minds ('cognitive distortion');
- By directly altering inputs and/or outputs (for example, working harder, in order to garner greater recognition);
- By leaving the organisation.

Whilst equity theory is helpful in trying to reach an understanding on the psychology behind worker motivation, it has garnered criticism in terms of practical application.

For example, workers are liable to ascribe differing personal value to their own contributions – meaning that two workers at the same level, doing the same jobs, for the same money, may feel entirely differently about the fairness of the situation. Alternatively, a worker may make an entirely erroneous assessment of those around them.

Reactions may also vary wildly. For example, feeling themselves over-compensated, one worker might respond by increasing their effort; whereas, another may respond by adjusting the value that they previously ascribed to their own personal inputs – which might result in a decrease in efforts.

In *Managing the Equity Factor: After All I've Done for You* (Huseman et al., 1987), behavioural psychologists Richard Huseman and John D. Hatfield point out that much of Adams' original research was conducted in laboratory settings – and, as a result, the applicability to real-world situations is questionable. What is less questionable, however, is that workers value fair treatment – and that a perceived lack of equality can be a considerable demotivator.

Leaders that want a happy and motivated workforce need to strive to maintain fairness in working relationships and focus on output rather than personal issues. Giving workers opportunities that enable them to grow their careers – either by learning new skills or providing valuable experiences – is another way to keep workers engaged.

This kind of personal development can be take myriad forms: from presenting opportunities for workers to network with other departments or interact with specialists to develop skills or acquire knowledge about a field of interest.

THE NAKED TRUTH: ENGAGEMENT

Whilst many organisations still operate traditional, hierarchical, command-and-control-style management structures, in recent years many have had pause for thought as they've witnessed their competition overtaking them. Panicked, they investigate – only to discover that – whilst products and services are comparable – the more collaborative structure of their competition's business model helps facilitate higher levels of worker engagement.

Worker engagement is vital for every successful organisation, as it creates greater empowerment and ensures full job-role participation. When workers are engaged, innovation naturally increases, productivity rises and the organisation will become more efficient. In short, the entire operation becomes infused with purpose, energy and enthusiasm.

Engaged workers are not only more productive and happier at work, but also work more safely.

Building positive engagement at work is all about connecting with people and in this chapter we've underlined the importance of respect, the value of acknowledgement of contribution, and considered ideas on motivation and the use of goal-setting.

Ultimately, engagement – like all people-related aspects of your business – needs to be fair and appropriate, creating a sense of equity that everyone can share in and enjoy.

GET NAKED: ENGAGEMENT

1. How would you rate the level of worker engagement at your organisation? What causes you to think this?
2. How often do you get recognition for 'a job well done'? Would you say that it's delivered more as praise, or as acknowledgement? When it happens, does it motivate you to work harder?
3. How engaged do you feel in your current role? What would make you feel more engaged? How can you share these feelings with your supervisor in a constructive way?
4. Think back to the last time you recognised someone's contribution to safety at work. Did you use praise or acknowledgement? What impact did this have on the person? How has this recognition changed or affected your relationship with the person?
5. Marciano's RESPECT model articulates seven principles that strongly influence worker engagement (Recognition, Empowerment, Supportive feed-back, Partnering, Expectations, Consideration and Trust). How well does

your organisation manage each of these? And now, thinking personally, where are the opportunities for yourself to focus now?

6 Look at the goals your organisation has for workplace safety. Do you feel that they are SMART (Specific, Measurable, Achievable, Realistic, Time-bound)? What do workers think?

Part 3

Looking forward

14 The safety practitioner

Safety practitioners today find themselves in a young profession that has witnessed significant and fast-paced change. The Institution of Occupational Safety & Health ('IOSH'), the chartered professional body for safety and health in the workplace, now has over 46,000 members though it has existed only since 1945.

In this chapter, we'll discuss the evolution of the role from safety officer to safety practitioner, and explore what the changing world of work means for those working in the profession today.

Despite its worthy aim of ensuring safe and healthy workplaces, for too long the profession has been dogged by negative stigma and low public perception, so we'll consider how to break this perspective and consider ways to reposition the role as 'safety leader' and add tangible, sustainable value to the organisation today.

Jock was a warehouseman, in his mid-fifties but still as fit as some of the younger lads on shift. Married 28 years, kids grown up now. Proud to serve his employer for almost 31 years, and well able to perform his daily duties without issue or error. Checking vehicles in and out of the depot, counting pallets, boxes and cases as they were unloaded from the trucks and onto the racks. Counting them out again as they were split down and loaded up for delivery to customers. Jock enjoyed his work, and –whilst by his own admission he was no Einstein – he took care to ensure his numbers always tallied up. Happy in his own world of Stock Keeping Units and Delivery Manifests, it was a steady life. Until it all changed.

One morning, as Jock arrived for work – 15 minutes before clocking on time, as usual, he never liked to be late – he was called to his Manager's office. "Jock, we need you to help us with something," was the opening line. Always happy to help, Jock's ears pricked up, and he waited to learn what was needed. "You see, it turns out we need to be watching what we're doing with all these forklift trucks and lorries, make sure our racking isn't going to fall down, see that no-one gets run over in the yard. You'll remember when Alan had his toes squashed by a forklift truck a few months back? Well, it seems that the regulator isn't that happy with us now. They reckon that we're not 'managing safely'. I told the guy that we'd only had a handful of accidents since we opened this depot five years ago, but he didn't seem impressed. Said we should not have any! Told me that there are laws about safety at work, and that if we didn't comply with them, somebody could be killed. And . . . hang on, what else did he say – oh yes, that I could even go to jail. Go to jail? Me? Well, it's not my fault if someone here doesn't watch what they're doing, is it now?"

Jock sat carefully listening to his boss. It seemed that this was an important issue for him.

"So, it turns out we should have someone making sure that we've ticked all the boxes. Crossed the t's and dotted the i's as it were . . ." continued the manager. "And everyone knows that you love that sort of thing Jock. I've always said that when it comes to checklists, you won't find better than old Jock here." 'It's true', Jock thought, 'I am pretty good with checklists. And spreadsheets . . .'

"So, Jock, what do you say? You're the man, right? Don't worry, it's not much more than you currently do, and you'll be able to fit it in around your other duties. What do you reckon?" Still flattered by his manager's recognition of his admin skills, Jock looked back from his daydream eagerly: "So, is this a promotion boss?"

"Well Jock, you know things are tight right now. We don't have a lot of spare cash going around. But do a good job and we'll see, hey? Tell you what, though, you do get a new title. You'll be our 'Safety Officer'. And we can send you on a course too if you like."

"But what about my current title?" quizzed Jock. "Easy," replied the manager. "You'll still be 'Team 2 Warehouse Controller', but you'll also get your name on the staff noticeboard as 'Safety Officer'." "Sounds important!" mused old Jock. "Yes, of course it is, but really it's just about getting the paperwork in order, so that if that guy from the regulator ever comes back, he'll not be able to send me off to jail . . . Send me off to jail! Ha! These guys in local government, they really do think a lot of themselves . . ." trailed the manager, as he wandered out of the room.

Jock stood there, thinking to himself: 'Jock Roberts, Safety Officer. Yep, sounds important all right,' and off he went to find a quiet corner of the warehouse to start drawing up his first 'safety checklist'.

Back to the future

It's 1975, a time of high political pressure, workplace democracy is shaken as storms brew in social as well as economic climates. A Safety Officer by the name of M. C. Bryant writes an article for the *Annals of Occupational Hygiene* entitled 'A Safety Officer's View of the Future'. Bryant's plea was strong, that in these times of great change, the UK, as a nation rich in experience must: "harness these skills to resolve the hazards which plague our environment. Our objective is clear – a safe and healthy future for everyone who works."

With words resonating through many developed economies around the globe, Bryant suggested that the role of the Safety Officer had unfortunately "developed in a fragmented and somewhat undisciplined way" pointing out that the common aspects of the job were typically to facilitate poster campaigns, act as a workplace hazard inspector or at times a safety policeman holding limited skill in proposing practical remedy to the risks he encounters.

Sure, we've all sniggered at the idea of a safety 'policeman', eagerly patrolling the workplace with his clipboard in hand, ready to pounce on offenders for not wearing their Personal Protective Equipment, for horsing around, or to conduct some sort of time and motion safety review, but almost four decades ago, back in 1975 it seems that this was how it was. Luckily, today it's entirely different. Isn't it?

Well, not really. Health & Safety has for too long been 'done to' people, by people just like our friend Jock. In the years since Bryant's plea a one-way relationship that revolves around checking that people are doing the right thing, complying with Standard Operating Procedures, and working safely has developed in many organisations. Communication from Safety Officers would occur where employees were found to be working unsafely, ignoring hazardous conditions, making mistakes or following an accident. Conversation starters would typically include:

- "I see you're not wearing your safety glasses."
- "You should be following the pedestrian walkways."
- "The law says we need to do this."
- "I'm going to do a safety audit on you."

Such assertions reek of the trusted responsibility that has been levied upon these individuals that has been naïvely undermined and eroded by their own approach. In such organisations, as a result, the ownership, action and accountability for safety all rest firmly on the shoulders of the safety officer and rarely moves beyond.

Many of these individuals relish the call to arms, the responsibility and having 'the power to call stop'. With a direct line to the Managing Director or CEO, and the final decision to block changes or halt the production process on grounds of safety, the Safety Officer prides himself on his ability to assert – and demonstrate, that 'safety is the first priority around here'. In the background though, life goes on, and the goodwill once freely given by peers on the management team rapidly dries up, allies become enemies, closing their ears, and cooperation turns to obstruction. As Bryant wrote back in 1975: "Many Safety Officers have suffered from the isolation imposed upon them by the organisation for which they work." Disappointingly, the truth is today many still do.

More than 40 years later, in many organisations around the world, the job of the Safety Officer can still be a lonely one. Divergent agendas are, sadly, all too often commonplace – where the Safety Officer wants safety first, ahead of production, profit and performance but the production manager wants to get the goods out the door. Or the Safety Officer wants to eliminate risk, but the Board want to take risks in order to develop new business models. Horns become locked and progress grinds to a halt. It's all too easy though to blame the individual here. We cannot fault the Safety Officers, those people who have been delegated these duties, who strive to only do their best. Indeed, where we have ended up is more a product of the organisational culture, that has discreetly *facilitated* this Catch 22 situation.

A tale of three practitioners: the monk, the missionary and the mercenary

With his 'promotion' to 'Safety Officer', Jock applies his particular skillset to devising a comprehensive health and safety policy, and an accompanying manual of detailed procedures. His belief is that 'Rules are rules' and if there isn't a rule for something, he's quickly able to write one and is ready to enforce it. With years of experience, drafting these rules is a straightforward process as he's familiar with the tasks and the work environment, so he quickly builds a safety manual that has the management team speechless.

The monk

Attendance at a training course generates a sense of self-worth that Jock hasn't felt before, and realising the importance of consultation, he diligently distributes drafts of each new procedure document to all the managers on site to seek their comments. When the negative feedback comes in, Jock views it as picky, and that the managers don't really understand safety the way he does, instead they seem more interested in avoiding cost, or the extra activity that Jock's new process requires. And so, it goes on, the careful preparation of procedures and the meticulous noting of their issue galvanising the importance of the role of Safety Officer until eventually Jock is successful in securing his own office. At the end of the warehouse, near the emergency exit, it's a bit of a walk to the coffee machine, the toilets and the other offices on site, but it does provide a good view onto the shop floor.

The procedures continue to be issued, though Jock has noticed that comments on his drafts have dried up. The managers have at last realised that he is the expert in this difficult area of workplace safety. But he's also noticed that several departments are slow to implement the new ways of working, and in some areas of the business they just seem to ignore the procedures altogether. 'Hmm, this just shows that the company doesn't really care about safety at all,' muses Jock. Nevertheless, he continues to maintain the networks with his new acquaintances in safety positions at other organisations, extracting their best practices to use back at his place of work in the desire to ensure that his procedures are always the very best they can be.

Jock has efficiently created a niche for himself as a Subject Matter Expert on an Apparently Exceedingly Difficult Specialism. His work is clear to him, to establish Best Practice in safety in all areas of the organisation. If he can do this, the accidents will stop, and the pay rise will come. What Jock does not see, however, is that he has effectively distanced himself from the rest of the organisation. Jock has become what Mike Buttolph of Cranfield University in the UK refers to as a safety *Monk* – devoted to the meticulous maintenance of the good book – his Safety Manual, living in isolation in his office at the end of the building (Buttolph, 1999). Whilst some of the team still believe he has some knowledge or skills that add value, others feel he's become unreachable and appears to be operating in a different world to them.

The mercenary

Buttolph asserts that the second form of practitioner is the Mercenary. A safety 'activist', who takes centre stage within the organisation when it comes to matters of safety. She's the go-to person and can usually recite the letter of the law for any given workplace risk, explaining in detail where the breach is, and what the penalty for failing to resolve it will be. Her training sessions to the management team revolve around how seriously safety failings are taken by the regulatory bodies, and ensuring that the managers are all too well aware that on their mission to 'safety first', she has been given carte blanche authority from the CEO to stop any task, machine or process that she feels is not being undertaken with safety as priority. Regarding herself as the 'protector' of the organisation, her 'command and control' style of influencing is tolerated by her colleagues, none of whom wish to fall foul of the law and end up on the receiving end of an enforcement order. Like monthly monies to the mafia, there's always someone willing to buy her coffee in the hope that it'll be remembered when the Safety Inspector arrives on site.

The Mercenary has successfully turned safety into a mystical black art, a sort of 'stealth & safety'. Under her instruction, her peers have all realised just how complex the science of safety is, and that there really is no-one who really understands how to carry out risk assessments, lead safety studies or get to the bottom of accident investigations when she's not around. Even when her peers have a go, these managers find themselves unable to cover off the broad range of risks that they know the Mercenary will find on her return, so the majority leave well alone – it's far better that the Safety Officer stays the fount of all knowledge and they get back to doing what they know best, running production.

Beyond her precise attention to detail, the Mercenary cultivates an air of nonchalance and can appear reluctant to get involved. Her terms of engagement are usually split between the assumption that no-one else is as skilled as she, or because no-one else has the desire to try to solve these complicated safety puzzles. Safety meetings are held with all the formality of a regimental inspection. To prove that safety is a line management responsibility, the production manager chairs the meeting, though with the Mercenary at his right hand to provide the direction, reference to the safety rules and regulations, and in times of hesitation, the words to fill the voids of his conversation.

The company prides itself on its safety management system, confident that between the covers of that pristine yellow lever-arch folder each manager has been given (and signed away their lives for) lies a wealth of detail, legislative reference and specification. The folders gleam on the shelves like the latest special offer carefully positioned on the used car forecourt. Updates to the policies and procedures within the folder are stamped 'urgent' and despatched from the safety office on a monthly basis. Upon receipt, the managers slide them into the polythene pocket at the front of the folder, mentally noting their plans to file them into exactly the right thumb-tabbed, indexed section before the next safety audit. *Ah*, the audit. Conducted each quarter in a hard-disciplined style that smacks of military inquisition, they leave auditees floundering to find evidence of compliance and demonstrate their good practices. The urban myth of another department attaining an audit score of 90 per cent continues to float through the air, and seems still potent enough to provide something to strive for.

Like the Monk, the Mercenary has also found herself a niche as a safety specialist. A tolerated arch-angel who springs in to save the day and protect the company when things go wrong and the threat of regulatory action becomes reality. The mystique with which she shrouds her skills swirls like a superhero's cape.

The missionary

The final stereotype in Buttolph's trilogy is the Safety Missionary. This pure and faithful evangelist is on a mission to convert us all to become believers in the power of safety.

A close call with a workplace accident some years earlier provided the calling he needed to lead himself towards the light. Adorned with the very latest in Personal Protective Equipment the Missionary always practices what he preaches.

Always looking on the bright side of life, the Missionary sees safety from a humanitarian perspective, knowing that his colleagues are intrinsically good people who want to work safely, and that the managers are constantly striving to always do the right thing – though on occasion he accepts that they are too overwhelmed by their day jobs to do it. The Missionary believes that these managers need someone to confess their safety sins to, to explain just why they couldn't get safety right and then to learn the lesson of what to do next.

Penance is rarely imposed. His parables are preached through the form of best practice examples, benchmarking opportunities and new safety checklists. All well-intentioned and gently-offered – but they leave managers with a feeling of disconnection, this off-the-shelf-one-size-fits-all just . . . well, just doesn't fit.

Rather than seeking clarity, the manager keeps his head down to avoid more of the same from the Missionary, and the workers pull together to protect their colleagues. Of course, it's hard to fail to respond to the Missionary's joy and passion as he smiles constantly on his audit tours of the workplace, so civil greetings are exchanged freely – but, despite the warm feeling and well wishes, the Missionary is out of touch with the realities of the shop-floor and the strategies of the boardroom.

Blind eyes, bureaucracy and niche-carving

Whilst Buttolph's portrayal of these safety stereotypes was developed nearly two decades ago, it still provides for an excellent and relevant analysis of the development of the role of the safety practitioner, and many organisations may feel a sense of familiarity with one or more of these characters. Each one shows how the safety profession has the power to – unintentionally – drive the management of safety in the wrong direction.

- The Monk has a dogmatic drive for his organisation to be the best in safety, but cannot see beyond his own spectacles. He has failed to engage any of his stakeholders and has become a slave unto his own agenda.
- The Mercenary builds bureaucracy through generating a sense of fear within the management population. Fear of enforcement, of production loss, of penalty. A fear against which only she can protect. Believing in the superhero's invincibility . . . but the smoke and mirrors don't help anyone.
- The Missionary's gentle character and genuine desire to be at one with the world prevents him from 'raising the game' – and driving a sense of pace and urgency in his modern business as they pursue performance improvement.

Each carve their own niches within the workplace, isolating themselves on a safety island, where management peers are happy to turn a blind eye and let them reside, so long as they don't have to do the 'safety work' themselves.

What went wrong?

There are two principal catalysts for the evolution of safety into Monk, Mercenary or Missionary roles. First, the delegation of 'all things safety' to a single person, the 'Safety Officer'. Historically, the exclusive remit of the Safety Officer was the prevention of traumatic injury. Why this focus only upon those most serious

REFLECTION POINT

The health and safety profession has long suffered challenges with public perception. Ideas of 'safety policeman' jump quickly to mind for many people.

Think about the safety professional in your business (this could be you!) and identify five words that describe this person, their style and their approach.

Work quickly and don't spend too much time thinking – just go with the words that pop into your mind.

____

____

____

Look back at these descriptors. Which of them do you feel are important to support your organisation's vision for safety? Draw a circle around those that you feel are necessary.

Now think about your company's vision for safety. What are the characteristics needed for a safety leader to turn this vision into reality? Which of these does your safety leader currently have? Are there some areas where focus is required now?

____

____

____

risks? If we consider the role of those managers within the production environment, on an average day, they face many challenges and problems: production pressures, customer complaints, deadlines, quality control and assurance, product recalls, plant and equipment breakdowns and maintenance, raw material shortages, cost of goods sold, staffing and absence to mention but a few. Some managers may have benefited from some general management training – however, many others will have learned their craft through the constant immersion is these daily events. But no matter which learning route has been followed, the rub is that the average manager is simply not trained to deal with danger.

Accidents, whilst certainly opportunities for learning, by their very nature (and either our aspiration or planning – or perhaps even luck) do not occur with such regularity as these other management issues. When they do arise, the manager strives to balance production and safety – keeping the machines running, whilst taking care of the injured party. Salient lessons from the accident will be observed, where possible, during the return to normal operations. Ordinary, routine matters of occupational health and safety, together with less severe hazards, may float by like leaves on a river.

In good organisations, the Safety Officer will be there, ready to catch these safety issues in his net, and resolve them with little fuss, as his own contribution to 'keeping the machines running'. But this approach does nothing to further manager's ability to manage safety. The Safety Officer, like the analogy, has become the organisation's safety net – and, so long as he keeps catching the issues, why bother doing anything different?

The second reason is the creation of the black art of bureaucracy. The ideology of a structured approach to safety through the sharing of information and open, cohesive working practices and relationships has been frequently obscured beneath layers of bureaucracy as well-meaning Safety Officers, unsure of expectations find themselves in a Catch 22 situation, just like Jock, trying to help management, but unaware of the need for – or application of – proportionate risk management. The officer, trying his best, diligently drafts his policies and procedures to fill the safety file and 'keep the regulators away'. Safety Officers have meticulously developed libraries of risk assessments, revised SOPs and at the same time, alienated management who see their time spent mired in paperwork which seems irrelevant to the works required and too bureaucratic.[1] Accordingly, the great risk we run is for organisations to assume that everyone within shares the same common understanding of 'how the organisation works', and specifically, 'how safety works'. Before long the organisation has developed its own problem-oriented culture towards workplace safety, and negative stigmatisation and isolation has taken hold.

Overlaid across these two challenges are fast-growing perceptions of an increased litigation culture in many developed countries around the globe which has fostered a fear-based philosophy where the production of such voluminous documentation has become regarded as the *only way* to provide vital armour-cladding to protect against legal action. Ironically, this approach furthers the disconnect between safety and reality and strengthens the belief that safety can only be 'done' by the experts. This self-protectionism has not gone unnoticed. In a speech to the Institute of Public Policy Research in May 2005, then UK Prime Minister Tony Blair warned that the UK had "a wholly disproportionate attitude to the risks we should expect to run as part of normal life". Disappointingly, it would appear that this 'disproportionate attitude' has continued in many developed nations, and in some, clear signals can be observed of it gathering momentum.

Some practitioners have been granted the moral fortitude and inner strength to accept their isolation, and see themselves as a misunderstood technical expert. Others may find themselves uncomfortable being so far out on a limb, and find ways to melt back into the team through a different position. Luckily, some – perhaps, sadly, a minority – experience an epiphany as they realise that times have changed, and they must change too . . .

So, is there a disconnect between what OSH professionals think they should do and what organisations think they should do? I suspect that there is not just a simple gap, but indeed there are *significant differences* between how senior leaders and safety practitioners perceive OSH is being managed within their organisations. Why? What caused this detachment?

Changes

Wouldn't it be great if we could point to one thing and confidently say: "there, that's what went wrong!"? Of course, it would! We could then efficiently fix the problem and forget al.l about the 'safety issue' until next time. But it's just not as simple as that. We have arrived at this point in our safety journey because several factors have collided.

The world has changed

Under the influence of shifting economic, social, political and demographic conditions the world of work in which we all operate has changed and will continue to constantly change. Corporate Social Responsibility, new technologies, broken-down borders and transient workers, international trade and economic partnerships developing in the Far East, Latin America and in Europe all bring new challenges that require new skills as globalisation picks up pace and changes the working world. Unfamiliar legal landscapes are discovered as organisations expand across the globe, presenting us with regulatory and litigious risks that may not resemble those found in home countries.

The work we do has changed

In many developed nations, the shift from industrial to service industries is palpable. As we begin the new 'knowledge economy revolution' we must realise that the work we do today, and where we do it, is radically different than it was in the past. We continue to outsource many activities as we return to focusing on our core product or service provision. As a result, supply chains become elongated and costs driven down as contracts are awarded to the lowest bidders, sometimes on our doorsteps, but more likely situated across the globe in developing nations. Disasters like the Rana Plaza building collapse in Bangladesh, which killed over 1,100 people and injured more than 2,500 in April 2013, have raised the bar with regard to human rights and social responsibility, reminding us of our moralistic duties and forever affecting the matching of task and worker.

The way we work has changed

The advent of the 'executive academic' graduating from renowned business schools such as Harvard, Kellogg, Duke, Wharton, London Business School, INSEAD, IMD and many more around the globe has had profound impact on the way we work. New models, principles and philosophies for business are enlightening boardrooms and literally turning business strategy on its head. As the knowledge economy picks up pace, organisations are faced with myriad consultants offering to find and fix problems they never knew they had – all in the name of business improvement. Such business changes render some habits and hardware obsolete whilst creating valuable opportunities for new ways of working and living to emerge.

The way we do safety has not changed

Risks today are no longer purely physical, related to traditional causes like industrial machinery. Yet, the way in which we do our work is not really changing. Whilst there may be a tangible shift in the language we use to describe our work – for example from 'compliance' to 'risk management', or from 'reporting' to 'governance' – we continue to operate with our traditional methods in a safety silo, distinct and detached from the rest of the organisation. With organisational cultures effectively promoting the evolution of Monks, Mercenaries and Missionaries we become stuck on a slowly-turning wheel, unable or unknowing how to make it stop.

These are great catalysts for reviewing how we think about, present and 'do' safety. But it seems common that when organisations find themselves on a performance plateau – having successively reduced workplace accidents through successive introduction of physical measures such as machinery guarding, administrative controls like training and supervision, and then attempts at influencing behavioural change through observation-based programmes – they find it all too easy to revert back to the old models of 'doing safety'. But what got us here in the past, won't get us where we want to be in the future. As Henry Ford said "If you keep doing what you're doing, you'll keep getting what you got."

Over the last 30 years the profession has experienced profound evolution, from the redefinition of roles and responsibilities, to a dramatically changing regulatory landscape. The approach of the safety practitioner must also continue to evolve as organisations and expectations change around us. The future will be for those who not only keep pace with change, but can manage to stay ahead of the curve.

To break through the glass ceiling and truly drive added value from safety back into the business we need a step-change in our approach, from being reactive to being *responsive*. This means we need to:

- STOP reacting to workplace hazards and START responding to the *output* of our risk management programmes;
- STOP reacting to accidents and START responsively *learning* from history to identify themes, trend and opportunities;
- STOP benefitting from safety 'by chance' and START *planning* to experience safety by *choice*;

- STOP doing safety activities based on the competency and availability of limited safety resources and START aligning our safety strategies with the organisation's vision, purpose, aspiration and needs;
- Ultimately, STOP relying on safety leadership based on availability of role models, or the CEO beginning speeches with safety and START building capable leadership based on being fit and ready for future safety and workplace risk challenges.

There is no room for the Monk, the Mercenary or the Missionary in the Safe New World. The role of the Safety Officer has evolved, and will continue to evolve, beyond the specialist practitioner remit, and into the safety leader.

The smart organisations are already there, moving safety to a core value and exploring how systems, leadership, engagement and motivation all influence workplace behaviours.

These organisations are repositioning their people – like Jock – from safety 'doers' to safety leaders and capitalising on the value that a *naked safety* approach brings to the organisation. From more engaged employees and increased wellbeing, to reduced risk, enhanced corporate reputation and better governance, there's 'money left on the table' and it's only right that we put it back where it belongs.

Revitalising objectives

So, what is the purpose of 'safety' within a business? Is it to assure legal compliance? To reduce accidents? To investigate what went wrong? To carry out audits? To crunch the numbers?

Well, in a sense it *is* all of these things, some of the time, but, as we move forward, if we truly desire to drive a step change and gain the additional worth that we've been leaving behind, we must re-think our purpose and start to consider how safety can contribute value. Here are five ways to strip back the approach to safety and bring the bounty back to the boardroom:

1. **Support the business**
   In order to enable productive change, safety needs to be an 'energy releaser' within the business rather than an 'energy drain'. A crystal-clear understanding of the organisation's mission, strategy and goals is of paramount importance if the safety practitioner is to be able to confidently present herself as a business partner that can support other managers' achievement of their objectives. If we can't show how safety supports and adds value to the business, just like the Monk and the Missionary, our services won't be utilised. Keep in mind that any approach to setting OSH policy needs to be evidence-based, relevant, practical and in direct response to the real risks of the workplace.
2. **Lead from the front**
   Naked safety goes beyond the demonstration of technical ability. On our journey to safety excellence, the safety practitioner must show the way

forward and act as the GPS, the satellite navigation system. In times of corporate change, it's fair to expect that resources may be thin on the ground, so the ability to lead through influence and persuasion is vital to build networks of influencers who can become advocates for safety throughout the business.

3 **Be collaborative**
In order to move away from the silos that the Monk, the Mercenary and the Missionary all created for themselves, safety needs to develop allies and build strategic partnerships across the business, infecting other departments and functions to co-create shared activities, goals and objectives. Building safety into the way the business works smashes down the silo walls and allows everyone to see how they contribute to the corporate vision. As organisations realise that the 'H' in Health & Safety is vitally important for consideration, practitioners need to get up to speed on key health issues, and adapt their skillsets. This doesn't necessarily mean becoming experts on workplace health, but more likely cultivating more efficient and effective partnership working with occupational health and hygiene specialists.

4 **Think like everyone else**
The time for thinking like a Safety Officer is over. In the *Safe New World*, the safety practitioner must be all things to all men, by operating as a miniature model of the organisation. To discover and enable the hidden added value, we must think and function like the entire business. Yes, the *entire* business. There is much to learn, but by engaging with peer groups we not only learn valuable insights to develop ideas for re-presenting the safety and risk approach, but also have perfect opportunities to join-the-dots and grow new partnerships. Get together with the movers and shakers in the other departments and see what makes them tick. When they strip things back, what are the *naked* factors that bring them results?

- **Think like Marketing:** Understand the needs and desires of our customers with the insight of the most creative marketeers. How do these guys create sexy and provocative campaigns?
- **Think like Research and New Product Development:** Develop tools, techniques and services that meet the needs and beat the desires of our customers, with the technical ability and nuance of the most innovative Research and New Product Development (R&NPD) engineers. Where does R&NPD get their new ideas from? How do they spot gaps in the market?
- **Think like Production**: Keep everything running smoothly, and delivered just in time. How does the Production team keep the machines running? How do they ensure that they create just enough of the right product to satisfy customer demand?
- **Think like Quality Assurance:** Incorporate robust assurance and verification processes to ensure everything hits the standard expected – and then beats it. What checks and controls deliver most confidence?

- **Think like Engineering**: Like the perfect Planned Preventive Maintenance schedule, we need to be able to 'find and fix' safety and risk issues before they arise and cause downtime. If we operate in *reactive* mode it will be too late to add the real value. How do the engineers set their schedules? What are the signs they notice that helps them stay ahead of the game?
- **Think like Sales:** Present products and services in a way that attracts attention, inspires action and generates demand for what we do. What's in the mind of the leading salesperson when they begin their pitch?
- **Think like Finance:** Build value into each and every one of our products and services that gets released when they are deployed in the business. How do the finance team measure value? What are the established methods for gauging return on investment?
- **Think like Customer Service:** Delivering products and services at the point – and time – of need in a way that satisfies customers' demands. How does this function continuously delight the customer? And when it goes wrong, how do they drive positive learning from the experience?

REFECTION POINT

If we can *think* like the rest of the business, then we can *serve* the rest of the business better and gain greater alignment. So, in this thought experiment, reflect on how each of the departments in your business function.

Strip back your thinking to identify what distinguishes each team from other departments. What are the *naked* factors that are the hallmarks of *how* they do their business?

Department	*Naked factor*
Marketing	
R&NPD	
Production	
Quality Assurance	
Engineering	
Sales	
Engineering	
Finance	
Customer Service	
Other department	
Other department	

5 **Deliver tangible value**
"Show me the money!" In the Hollywood blockbuster *Jerry McGuire*, Tom Cruise's character epitomised everyone's favourite love-to-hate capitalist. But let's be clear. Business does business to make profit. Now, it's not about turning the Monk into *Gordon Gecko*, but in order to be aligned, safety needs to be able to either reduce the cost of doing business or actively contribute to the bottom line as an income-generating stream. Traditionally, safety has demonstrated how reducing accidents can effectively 'save money' for the business, but this is always based on hypothetical standardised data or involves a serious round of *reactive* internal data-trawling and number-crunching. Moving forward we need to be much clearer about how the whole range of occupational health, wellbeing, safety and risk management activities provide solid ROI (Return On Investment) and support the financial health of the business by using Cost Benefit Analysis (CBA) to help show the impact of projects, programmes and strategic activities. Yes, it's a sensitive area talking about personal injury and money in the same sentence, but unless we do, we'll never see the true value that safety brings to the table.

Remember that beyond reducing costs related to absenteeism, accidents and disease, OSH measures also support improvement of corporate image, reputation, position in the labour market and customer satisfaction. Can also reduce employee turnover and increase productivity.

So, with a clearer sense of purpose, and stronger alignment to the organisation's vision and goals, where do we go? What are the key attributes for safety leadership as we move forward?

Many practitioners already have the core technical skills and expertise to manage workplace safety and risk, but they don't always possess the softer skills, such as influencing and leadership, to engage with both senior decision-makers, budget holders and employees at every level in the organisational hierarchy.

I have a dream

We must move beyond the reactive safety approach, where the focus is on reducing accidents, saving insurance premiums and setting annual targets for percentage point reductions in accident statistics. For sustainable change to occur, we need to look to the future, envision what success looks like, and then engineer the route to reach it. Without vision, encouraging people to work together in the same direction – especially on safety – will be difficult, if not downright impossible. Now, before we dismiss the idea of the safety practitioner as visionary, let's just reconsider. *Naked Safety* is about stripping things back. If we keep peeling back the layers of the organisation, we find that safety is just another business imperative. Just another business activity that needs to be done efficiently and effectively. Just like anything else.[2]

So, if we consider safety in this way, then why should our approach to safety be different to anything else we do in the business? A brightly-visioned approach is crucial to organisational success. Think of some of the most successful companies in the world – Apple, Pepsi, Google, Nike, Mercedes-Benz – none of them arrived at where they are today by chance, through compliance, through simply reducing costs or just improving performance. Each one of these organisations, and many more that you can think of, reached their market leadership positions because someone had a vision of what they wanted to achieve. No matter where we look, whether the commercial world and Steve Jobs of Apple and Indra Nooyi of PepsiCo, or the likes of Nelson Mandela, Mahatma Gandhi, Martin Luther King or modern-day visionaries like Mary Robinson, the UN High Commissioner for Human Rights. Each just one person. Yet, each one had a vision that not only drove success, but a vision that could be understood by people. And those people followed these leaders on their journeys.

Corinne McLaughlin and Gordon Davidson suggest that visionary leaders are:

> the builders of a new dawn, working with imagination, insight, and boldness. They present a challenge that calls forth the best in people and brings them together around a shared sense of purpose. They work with the power of intentionality and alignment with a higher purpose. Their eyes are on the horizon, not just on the near at hand. They are social innovators and change agents, seeing the big picture and thinking strategically.[3]

McLaughlin and Davidson got it right, and if we deconstruct their thinking, we begin to see the game plan.

Change the game

As a profession, safety has suffered a crisis of credibility in recent years. The media around the globe has plagued those who strive to make the world safer for everyone. Stories like local authorities banning flowers in baskets hanging overhead, schools demanding children wear safety goggles to play conkers, to circus trapeze artists being required to wear hard hats have lit the match and then fanned the flames to create such a negative stigma that despite the constant rebuttals from professional bodies such as by IOSH, the International Institute of Risk & Safety Management, the American Society of Safety Engineers, the Safety Institute of Australia, the NSC, the British Safety Council, RoSPA and many other international bodies the role of 'safety professional' continues to be met with derision and muffled laughter. It's not surprising. Traditionally, OSH practitioners have rarely had *really* good news to share, publicising data on the number of accidents that have occurred, the people hurt at work, or made sick by what they do.

The time has come to change the game.

It's time to detox Jock.

We must focus our 'eyes on the horizon' – and write the future of safety.

Safety practitioners are the ones who hold the responsibility for how others see them and must become 'change agents', 'thought leaders' and 'value-contributors' for the organisation. How? By seeing the 'bigger picture' and developing a range of modern business skills that respond to the need for agility, decisiveness and 'thinking strategically' we can introduce new perspectives through demonstration of the ability to swiftly shift gears from grass roots to high level, adjusting language and content to localised needs and context whilst being sensitive to the cultural diversities in the ever-expanding world of work.

The practitioner of the future will hold the intellectual rigour to understand what's needed, translate vision and values into practical systems and common language, and lead a diverse range of projects that engage and influence a broad range of stakeholders through direct and indirect means to deliver added value for the business. Rather than orienting themselves around workplace problems, safety leaders will be focused on a higher purpose: facilitating organisational understanding of the safety issues it faces, and the likely impact of these upon both safety and financial performance; generating innovative solutions to lead sustainable change; and challenging the established, accepted norms, customs and practices when it comes to safety.

As a true leader, the safety practitioner should be an enabler of change, not a brake slowing progress. The *naked safety* practitioner is a discoverer of opportunity and developer of a range of useful, cost-effective solutions to de-risk business initiatives, rather than curtail them on health and safety grounds.

Convergence – going with the flow

We can acknowledge that at times the safety practitioner's agenda will naturally differ to those around him, but when common ground can be identified, it provides a solid foundation from which to build trust, collaboration and progress. Look to the corporate values to find direction on where the 'shared sense of purpose' or convergence can be found. In theory, and, we hope, in practice for many great organisations, the values will be well articulated and understood by everyone, and despite an element of natural personal interpretation, there should be sufficient to grasp and use as a starting point for building a convergent agenda. Such convergence or 'flow' – where all parties seek common ground and agreement – may be hard to find at times, but look deep enough and you'll find the golden thread.

This isn't about simply trying to overlay the safety agenda across the stratagem of other departments, functions or stakeholders, but in fact finding ways of tying them all together. Think of the way streams and tributaries converge into the larger river, creek or canal. Each does so using its own unique course, making its way past obstacles in its path, before adding its volume and value to the main channel.

Hallmarked by a shared desire for continuous improvement across safety, production and process (and other disciplines), convergent leaders share lessons openly but are also considerately mindful of not exposing the organisation and damaging its reputation. Opportunities for development are identified and grasped whilst concurrently risks are assessed, mitigated and controlled. In this way, efforts in safety become cumulative, rather than isolated, and the foundations for a solution-oriented culture are laid squarely in place. The creation of convergent agendas will build resilience into the organisational processes and help retain the continuous integrity of the working environment.

Adjust the mind-set

Many great organisations will already have fostered a step change in attitude toward safety, moving the remit of the safety practitioner away from 'a person whose job it is to cover all the weaknesses,' to a guiding role – acting more like a GPS or satellite navigation system to help the business achieve its aims. This is a crucial first step, for as long as the organisation views the practitioner as the 'person who does safety', progression will be difficult to achieve.

Safety shouldn't be the domain of one person, a team or a corporate department. We're all responsible for safety, so why don't we share the accountability?

As we saw in the chapter on culture, symbolism is key to shaping mindsets as symbols provide something upon which to fix our minds and cultivate a sense of belonging. Consider how safety is perceived currently in your organisation. What do the symbols tell you? Is it a vibrant subject, considered as an exciting part of the work activity? Do people strive to join the department or get involved in safety-related projects?

Can you identify gaps between current symbols, your understanding or meaning of them, and the corporate vision? It's fair to say that if you have found gaps, such disconnects may exist for others too. Instead of jumping to repair the holes, perhaps there's an opportunity to clarify the importance of these symbols, or introduce new, more meaningful symbols that provide a stronger degree of resonance and can more clearly be seen to support the vision.

Engage others

As we said before, safety isn't something you can *do* to people. It's something that can only be achieved *with* and *through* people. The most effective way to change the game and help refocus minds is by engaging others in the actual management of safety. Not just the *doing* of the risk assessments, or accident investigations, but actually creating ideas; building activities and campaigns; designing symbols, logos and brands; shaping the strategy; working out performance indicators and so on.

Involving workers is a major factor that directly influences the success of cultural change programmes in workplace OSH and risk management. Through participation employees can more fully understand the rationale for change and, crucially, feel connected to it, as part of the team guiding the business forward. Participation and deep involvement like this breeds ownership. The more people we can involve in changing the game, the more mind-sets we positively influence, the more awareness and understanding we raise, and, in turn increase the likelihood of success. Remember, it's all about behaviour. Changing cultures through building relationships requires effective field time: talking, walking, sharing, involving, engaging, modelling. It's time to get out there!

Developing competence

No matter what we do in life, there will be those who are competent, and those who are . . . *well*, not so competent. But what do we really mean when we say that someone is 'competent'? That they have years of experience? They hold a qualification? That they get things done? That they excel?

In safety circles, we've been caught up for too long on a legislative definition such as the one found in the UK's Management of Health and Safety at Work Regulations: "a person shall be defined as competent . . . where he has sufficient training, experience, knowledge or other qualifications . . ."

But knowledge, training and experience, in themselves, are not sufficient unless they are used efficiently and effectively. It's not about who get things done, or even sometimes, how, things get done, rather that it's more about being sure that the things that need to get done, get done, at the right time, and are done well.

In recent years, there has been an increase in the focus by the courts on competence in health and safety. A quick review of the United Kingdom's court reports will quickly reveal plenty of cases where both the employer and the safety practitioner have been prosecuted for lack of competence.[4] Whilst modern regulation around the globe generally requires a safety practitioner to be 'competent', it rarely defines what this means. The world's leading body for health and safety professionals, the Institution of Occupational Safety and Health (IOSH) issued a Code of Conduct to their membership which defines competence as: "a combination of knowledge, skills, experience and recognition of the limits of your capabilities".

Conspicuously, IOSH omit the inclusion of 'training' in their definition. Perhaps because training is the input, and knowledge and skills the output. Makes sense.

But is that how organisations view competence? Often, it's not. Isn't it more likely that an organisation, in the process of hiring a safety professional, will enquire as to whether a certain qualification is held? Yet a qualification is not proof of competence, it's merely evidence of attending or completing training. This is where the IOSH definition pulls it all together. With the incorporation of the final element, the Institution highlights the importance of responsibility, of knowing oneself and one's ability. A crucial element, often missed by the Monk, the Mercenary and, sometimes, even the Missionary.

Yet, each attempt so far has focused only on the gaining of ability. Perhaps it may be useful to by looking beyond the safety profession. Here's a much-used version from the clinical world: "The understanding of knowledge, technical and communication skills and the ability to problem solve through the use of judgement" (Norman, 1985).

Recent research in safety points to a growing emphasis on competent leadership, essentially the *application* of skill and knowledge, and robust decision-making as key factors in generating safety in the workplace. The competence, and therefore credibility, of safety professionals becomes a key influence on developing an organisational culture of safety, but we must be mindful that 'experts' can often perceive risks in a different way to the general public, in that they see risks potentially as smaller or less significant.

Safety skills

The National Examination Board in Occupational Safety and Health (NEBOSH) National General Certificate and National Diploma qualifications have traditionally provided an excellent (and highly popular) route for many practitioners to gain the knowledge base needed to begin a career in workplace safety. In the last decade or so, we have begun to see universities delving into the world of safety too, with many undergraduate degree and diploma programmes springing up around the globe. Postgraduate studies in safety have also become popular in recent years, each adding further credence to the import and value of the profession.

But we must go further. No matter which profession we consider, beyond the technical knowledge required, competence in its truest sense must also incorporate a range of soft skills. Let's go back to Bryant and his predictions for the future: "The safety professional will need to draw upon knowledge of both the physical and social sciences. He will need man-management skills, qualities of leadership and authority, and knowledge of the principles and theory of business and government."

Bryant points out, quite rightly, a selection of additional skills that are necessary – and I'd argue not only for a Safety Officer or practitioner, but indeed for effective safety leadership in general. Whilst the skills Bryant points to are certainly relevant, in today's modern business world, there are other attributes to add to Bryant's list, including analytical and diagnostic skills; influencing and negotiating; critical reasoning; creative thinking; problem solving; project management; customer focus; leadership and delegation; emergency response and crisis management; and a whole gamut of communication skills.

The performance paradox

Herein lies the dilemma. No matter which of the traditional routes to safety competence we look at – qualification-based or experience-driven – we find a significant void staring back at us. Whether degree, diploma or certificate – OSH training courses have not taught the safety practitioner the skills of management and leadership. Whilst technical competence may have been gained, the softer skills have not yet been seeded, let al.one allowed to germinate and grow.

There are three important aspects to competence – the first is clearly in its acquisition. The second, is in its application. But the third can be easily overlooked – the continual maintenance of competence. In the stereotypical Monk and Mercenary types of practitioner we observed these individuals feeling confident in their own abilities as Subject Matter Experts. For them, developing further their competence is unnecessary – they hold all of the keys to the safety problems they face. And after all, it's highly unlikely that anyone else within the organisation will have the desire to take their job. They have each successfully turned OSH into either a black art, or a subject so dry and difficult that it would be rare to draw interest or raise questions of their ability. By contrast, the Missionary recognises the importance of looking beyond his own nose. Through his ongoing benchmarking and acquisition of best practices from others he is, perhaps unaware, building and refining his own competence through these new ideas and practices, an essential action for all involved in the improvement of workplace safety.

Most of the institutional organisations for safety practitioners around the globe now include the maintenance of professional skills as a requisite part of ongoing membership. Such continuing professional development (CPD) schemes serve to facilitate a degree of continuous improvement through the identification of gaps in skillsets, or opportunities to develop further. Yet it remains the duty of the practitioners themselves to undertake such learning.

Ten traps for safety excellence

Safety practitioners of the future will certainly possess technical OSH skills, though it's likely that general management and leadership skills will dominate their skill set. Skills including developing the business case, Cost-Benefit Analyses, influencing, negotiating, coaching, engaging, selling, continuous improvement and people management will become more and more vital on the journey to safety excellence.

Though like any journey there may be obstacles that stand in the way of progress. Practitioners will need to think differently about how to plan to avoid these in order to contribute to the success of the organisation. Over the last 20 years I've worked with many safety practitioners around the world including plenty of monks, missionaries and mercenaries. I've been alongside old Jock and his contemporaries, and met visionaries who are keen to change the game. As a result, I've identified ten traps that can stop progress in its tracks for those intent of building safety excellence in their organisations. Let's look at these traps now.

The rise and fall of the safety department

I recall attending a recent safety conference where Judith Hackitt, then Chair of the British workplace regulator the Health & Safety Executive declared that the main aim of the modern safety practitioner is to do himself out of a job. The sharp intake of breath for many in the room was palpable. Many attendees wondered if she had gone crazy. Without the Safety Officer in role, wouldn't workplace health and safety issues just get ignored by organisational management? Surely the practitioner is needed with his safety net?

But, in fact, Hackitt was suggesting that the time has come for step change in how we 'do' safety and manage risks at work. A time for changing the way we 'protect humans from themselves' and safeguard organisations from injury and mishap. 'Making themselves redundant' must be the most noble aim a safety practitioner could possibly have. To have contributed so completely to the world that accidents and injuries no longer occur.

But if we think carefully about Hackitt's message, we can take our cue from her as to the recipe for success. Time and time again, the regulator of one of the safest countries in the world points to just one key ingredient: engagement. This may appear counter-intuitive. First, we're told that we must do ourselves out of our jobs, and then we're told that we need to engage people. Aren't these actions at odds with each other? Indeed not, they fit together on a path to maturity, and in fact, when we consider any of the safety culture models referred to in this book we can see them clearly reflected.

But how would practitioners know when their roles are surplus to requirement? When everyone is so fully engaged that the responsibility for safety has become shared: no longer as a burden to business – but shared as a value, a common habit, a privilege to be proud of. This is the end game. This, the most valiant aim of all, is what we, as practitioners or as business leaders, must strive for. The deconstruction and devolution of the safety department.

Globalisation, internationalised supply chains, shifting government agendas and relationships with regulators, growing interest of stakeholders and broader society in risk and assurance are just some of the drivers that will determine the future of the profession.

As we move forward, the acid test will be about clarifying the measurable value that safety contributes to support the long term financial health and prosperity of the organisation. To do this effectively we need to turn our attention away from fighting fires and calming compliance fears to building bottom-line business value.

Releasing energy, enabling change

As this chapter began, with the story of Jock, traditionally the safety role in many organisations is given to those at the end of their career, just to keep an eye on things as they wind down to retirement.

Over the last decade, however, forward-thinking companies are using safety as either a stepping stone to career development – especially in graduate programmes – or also as part of preparation for senior leadership positions. It's increasingly recognised that the majority of the work that safety practitioners do is about solid general management skills. These can be distilled to six key attributes that will underpin the success of safety leadership – whether delivered by a safety practitioner or a business leader – in the context of organisational safety, and beyond.

- **Visionary leadership** – the ability not just to see the future, but to be able to articulate it in a persuasive manner that inspires action
- The courage and confidence to **change the game** through constructive challenge, the use of clear and relevant language and useful innovation

Table 14.1

Threat	*Impact*	*Solution*
Symptom-focused approach	Focusing on symptoms restricts the organisation in maturing its safety culture	Dig deeper to identify the real root causes of accidents and the general approach to safety
Selfish silos	Safety is deemed to be the domain and duty only of the practitioner, preventing true ownership to be shared throughout the organisation	Create cross-cutting safety objectives that span all departments and levels in the organisational hierarchy
Lack of strategic direction	Well-intentioned programmes which lack rationale for implementation or are not in alignment with broader business objectives	Start with why safety is important to the business. Take the over-arching corporate vision and aims and align safety to these using similar language and terminology
Delegated ownership	The H&S department is seen as the sole owner of the discipline and the 'go-to' for each and every safety	Break down the barriers to leaders understanding what safety really is and create objectives for every department head that encourages them to lead with safety
Compliance-driven approach	A minimum standards approach to workplace safety leaves money on the table	Real value is only added when we move beyond legal compliance. Calculate and discuss the real Return on Investment for workplace safety by considering the additional revenue required to effectively pay for current accident rates
Reputation	Traditional approaches to safety have caused negative stigma to be permanently attached to the profession	Changing the game is the only way to change the perceptions. Look at how safety is presented and 'sold' within the organisation. How can you make it more appealing?

Continued . . .

Table 14.1 continued

Fire-fighting approach	Applying the quick fix band-aid to workplace OSH problems or jumping to improve safety after each accident is a slow road to improvement	Identify the underlying factors that create the environment where accidents occur and focus on these
New risks	A dogmatic focus can cause practitioners to get stuck trying to apply traditional risk assessment processes to radical new risks at best, or, at worst, see practitioners with their heads in the sand for fear of being unable to tackle something different.	New risks – such as psychosocial, workplace violence, transient labour, occupational road risk, ageing workers and nanotechnologies all bring new dimensions to safety at work which require new knowledge and understanding. Look beyond the traditional safety space for new ways of thinking
The Zero Trap	A constant focus on reducing accident or injury rates encourages the risk of ignoring deeper process problems	Encourage business leaders to see that a safe workplace is an output of effective safety management and risk control and build a balanced scorecard of leading and lagging metrics that measure what you value rather than value what you measure
Regulatory respite	The advent of regulatory bodies around the globe taking a 'partnership' approach with business will mean less frequent inspection, less stringent enforcement and ultimately less organisational impact	Use the change in enforcement approach to encourage ownership at local level and get away from wielding the threat of prosecution to empower managers to see safety as a positive enabler of business success

- Identifying a shared sense of purpose and bringing agendas together with **convergence** towards the vision
- Helping people to **adjust** their **mindsets**, through creating meaningful symbolism that allows people to feel part of the bigger picture
- **Engaging others** in the broader scope of safety, involving people, generating action
- Developing **competence** by building skills and knowledge and actively identifying opportunities for growth

When the practitioner gets these six attributes just right, they become not just good managers or leaders, but they release energy and enable positive change for the organisation, in safety and right across the business.

THE NAKED TRUTH: THE SAFETY PRACTITIONER

The languid historical delegation of responsibility to the Safety Officer has generated Monks, Mercenaries, Missionaries and Policemen who have become trapped in their own Catch 22 dilemma.

The world has changed. So too has the work safety practitioners do, and the way they do it. But a stronger shift in perspective, to take a more responsive view of safety, is required if we are to dislodge the negative perceptions and stigma which dogs the profession.

The role of the Safety Officer is redundant in modern business. Visionary leadership underpinned by a broad range of competencies is required to drive a sustainable step change in safety that releases energy, enables change and delivers tangible value. In this way, the persistent and negative stereotypes of safety practitioners can be broken down through 'can do' attitudes and real courage, conviction and passion to serve departments and functions within the organisation as true customers.

Today's safety practitioner operates in a continuously changing world of work. The traditional approach to safety, illustrated by our story of old Jock, is rapidly losing its effectiveness.

It's time for a New World Order. The power of safety must shift from singular to shared ownership. Deconstruct and dismantle the Health & Safety Department! Practitioners keen to engage others in safety improvement need to avoid over-complication and focus on the distillation of crucial information. Stop talking about legal obligation and instead demonstrate how OSH can be an 'energy releaser' not an 'energy drain' for the business.

Tapping into the thought patterns and working styles of other departments within the organisation can reveal fascinating insights that can help to build convergent agendas where everyone benefits.

The true purpose of safety is to support the business by leading from the front, building collaborative partnerships with convergent agendas that drive the business closer towards its vision, engaging and empowering everyone to contribute to the shared successes of a safe workplace.

GET NAKED: THE SAFETY PRACTITIONER

1. How is safety perceived in your organisation – as an 'energy releaser' or an 'energy drain'? Where are the opportunities to change the game to enable more positive action?
2. What is the approach taken to providing safety support in your organization – is it reactive or responsive? Look back at the list of five 'stop/start' ideas to explore how you might change the game.
3. How do you ensure that you develop well-rounded competence in safety that spans the building of knowledge, skills and experience to a level that is can be applied in an informed, useful way? What processes do you have in place to ensure competence is maintained and continuously developed?
4. Of the 10 threats facing the future of safety, which present the strongest challenge to your business? What steps do you need to take now to avoid these threats?
5. Recall the *naked factors* that you identified in the other departments in your organization. Which of these could you learn from and leverage to help change the perceptions of safety in your business?
6. There can often be a disconnect between how safety practitioners and senior leaders perceive safety is managed within their organisation. What processes do you have in place to ensure everyone is on the same page?
7. How can you share safety out across your business? What do you need to do now to help leaders take an active role?
8. Look back to the six key attributes for safety leaders. What are the opportunities for your own personal development?

Notes

1. See, e.g. Mulholland, R.E, Sheel, A.G. and Groat, S. (2005) Research Report 306, Sudbury: HSE Books.
2. Now hold on a moment, I'm not trying to provoke responses of 'hang on, it's all about the people . . .' but rather encourage you to see that safety isn't The Most Important Thing but instead just part of how organisations work.
3. www.visionarylead.org/articles/vislead.htm.
4. See especially the cases involving Prior Scientific Instruments in 2013; Perryman Properties 2010; George Farrar Quarries 2009.

15 Performance measurement

Performance measurement is an integral part of the effective management of occupational safety and workplace risk. All too often, though, we allow our metrics to revolve around the number of accidents or injuries that have occurred. This chapter will explain why this naïve approach can lead to the generation of a false sense of security and misguide us on our journey to safety excellence.

Leading and Lagging Indicators will be discussed – for safety and for health – before we explore a range of other ways of measuring the impact of our efforts to improve workplace occupational health and safety.

We'll mull over the real purpose and function of performance measurement and then work through the six steps of setting effective performance indicators, so that by the end of this chapter, we've completed a 360 and you're ready to consider how you can strip things back and measure safety and risk meaningfully in your own workplace.

Anyone who has every listened to classical music will appreciate that it's the whole orchestra that makes the music, rather than just the work of the violins, percussion or brass. Listen to a piece by the magnificent Austrian composer Mahler and you cannot fail to be excited by his emotionally charged symphonies. Throughout his work, Mahler appreciated the power of performance. He knew that to measure the value and impact of his sound, he must look broadly at the orchestral activity, not focus exclusively on repeatedly hitting specific keys or chords.

How does the leadership team of your organisation measure health and safety? Do you look at the whole orchestra, or do you focus on the notes? Chances are, the main indicator you will use is the number of accidents you have sustained over a period of time. Whether using actual numbers, or calculating accident frequency rates, it's fair to say that most organisations will present this data as evidence of 'good' performance. But the correlation between Lagging Indicators and actual levels of performance may not always ring true.

Measuring performance

The quality movement spurs us to want to measure safety at work. It's evolution, especially through the 1950s–1970s, has allowed us to get to grips with measuring almost all aspects of a production or manufacturing process. The work of people like Deming and Taguchi has had colossal impact on maximizing efficiency and controlling waste for many organisations, and it would appear sensible to apply the same logic to occupational safety.

We primarily seek to measure health and safety performance in order to provide robust information on the status and progress of our activities, strategies and processes in place within the organisation to identify, control and manage workplace risks. Measuring safety and health performance data can be useful to:

- raise awareness;
- accurately monitor how the organisation is performing;
- compare performance to other organisations within and beyond industry sectors;
- galvanise attention towards specific issues or areas;
- build motivation around OSH campaigns and initiatives;
- educate workers on previous events in order to prevent recurrence.

Performance measurement is a vital component of any management system and lies at the heart of every continuous improvement process. Used effectively, it can guide direction and focus efforts, helping us to form confident views on where we are with regard to overall aims and objectives, assist in the setting of priorities for action, and confirming levels of risk exposure. But often an over-reliance on one particular metric can lead us into a sense of false security around risk control, compromising the efficacy of our system and misleading management towards taking ill-informed decisions and actions.

It should be borne in mind that performance indicators are like shoes – one size won't fit everyone. But beyond the size, the colour, style, shape and much more in addition may not be to your taste. What works for one organisation may not work for the next. Taking a step further, indicators need to reflect the actual activity undertaken – you wouldn't for example wear a pair of Jimmy Choo's on your trek up Kilimanjaro now – so careful thought about what you seek to measure is vital before setting the metrics to gauge performance. Before you make your choice, it's prudent to check:

- Is the measure actually measuring what you want it to? (i.e. are cause and effect linked?)
- What behaviours will the measure encourage?
- Are these behaviours desirable?

> Measurement is the first step that leads to control and eventually to improvement.
>
> If you can't measure something, you can't understand it.
> If you can't understand it, you can't control it.
> If you can't control it, you can't improve it.
>
> Harrington (1991)

Lagging indicators

Lagging Indicators are widely used to measure safety performance, primarily because they are relatively straightforward in terms of data collection, and data analysis. As a result of this, Lagging Indicators have become common currency within organisations and also form part of many external verification, measurement and benchmarking programmes.

There are many examples of Lagging Indicators, including for example:

- injury frequency rates;
- injury severity rates;
- ratios between injury types;
- number of days lost per injury;
- sickness absence rate (%);
- audit scores.

Some Indicators may be based on lagging information but present data that can be useful to lead forward, such as OSH climate surveys, opinion surveys and employee engagement or satisfaction rates. We'll touch more on these elsewhere in this chapter.

Standard reporting using Lagging Indicators typically includes Accident (or 'Incident') Frequency Rates for example, Lost Time Injury Frequency Rates or 'LTIFR', and these can be useful for comparison with other departments or organisations. These rates show the number of incidents that occur for a given period of hours worked. A popular formula for this calculation is:

LTIFR = (# LTI/ # Hours Worked) × Standardised Hours (e.g. 100,000)

One hundred thousand hours is often used as the standardisation as it represents the equivalent of a person's working lifetime, the output of this calculation then can be used effectively to show the 'chances of having an accident' to workers in any given location. For example, an LTIFR of 0.5 would indicate that there is a 50 per cent chance that a worker will suffer a serious injury during their career. Expressing data in this way can often be meaningful – and persuasive – both at boardroom and shop-floor levels – and could be useful in early stages of cultural maturity development.

A point to note here for organisations utilising incident rates as benchmarking metrics is the importance of verifying the calculation convention used. In the United States, or for American-owned organisations, the calculation of accidents per 200,000 hours worked is a common convention – representing the equivalent of 100 people working for one year, and in many power and process industries, the calculation may be based on accidents per million hours worked. Benchmarking accident rates without careful verification of which calculation is being used may result in a skew by a factor of ten, and make a strong organisation look poor by comparison.

Incident Frequency Rates can be useful metrics as the function of the standardised hours in the calculation serves to make them resilient to organisational changes such as productivity, increased (or decreased) production hours, increase or reduction in headcount, organisational growth, asset acquisition or disposal and seasonal variations. But such rates provide 'smoothed out' or averaged over time, accordingly they may lull us into that false sense of security as a 'decent average' appears in our data.

Whilst they can add some value, we must bear in mind that Lagging Indicators *reactively* monitor the performance of risk controls and, in this sense, are a bit like the rear-view mirror in your car – helpful in giving a perspective on *what's occurred* as you have progressed along your journey, but not much use in deciding where to go next, or identifying what bumps, twists and turns you may encounter on your route. Whether driving on your journey – literally or metaphorically speaking – and only looking behind you may not be the best (or safest) method!

Due to their nature of being relatively easy to collect and reflect back into the business, Lagging Indicators are widely used, and perhaps, over-relied upon. We must bear in mind that Lagging Indicators – from an organisational health and safety perspective – are 'failure data', showing us what has gone wrong. It's at this point that we recognise another similarity with our peers in quality management – injuries are like quality defects, they are an output of a process fault. What's worse is that we continue to be seduced by statistics, believing that the absence of injuries somehow proves the existence of safety. But nothing could be further from the truth. *Safety* is really all about risk – how likely it is that people will be injured in our workplace. Looking only at events in the past, especially infrequent event such as accidents in a specific workplace will rarely provide either the clarity or a sound foundation for reducing risk.

As their name suggests, 'lagging' indicators may hold a delay between the actions we take and the output that results. By the time we receive a signal from Lagging Indicators it may be (and often is) too late to do anything to prevent the performance deterioration – the use of Injury Frequency Rates as a lagging indicator is a clear example of this. Reliance on Lagging Indicators can highlight a sizeable disconnect between aspiration and reality when we measure something that *doesn't happen* to us as proof that something exists.

Absence does not equal existence

Let's step away from accident rates for just a moment to explore this point in more detail by considering the 'H' of 'OSH'.

How do you define 'health'? Is health the absence of observable infection or disease? If you don't see the need to have a medical check-up, feel good and there are no visual signs of anything wrong, does that mean that you are 'healthy'?

Defining health in this way clearly makes no sense, yet we appear to continue merrily with such an approach when we measure safety. Achieving 'zero injuries' across a specific time-period does not prove the effectiveness of our risk controls, or, in simple terms the 'existence of safety'. This, we know, is common sense, as we have all, at some point in our lives, been unsafe and managed to escape without being hurt.

REFLECTION POINT: ARE YOU SAFE?

Think about an activity you do often – such as driving a car or riding a bicycle. Have you ever driven above the speed limit? Cycled without wearing a helmet? On each of these occasions have you suffered an accident? Does that mean that you were driving or cycling *safely*?

Planning to fail, or failing to plan?

Planning safety performance via the use of frequency rates as performance metric data is often done each year, in the boardroom, as the leadership team set out their agenda for the year.

Typically, the planning process will go something like: "What was our accident rate last year?" "Okay, let's go for a 10 per cent improvement this year." And then the reduction target is presented to the Safety Department for attainment. Right?

The Safety Team then scurries to implement new initiatives and monitor the rates in order to return the necessary reduction, in time for the revision of the target in about 12 months' time.

With safety, we isolate ourselves by operating in an ironically inverse situation. There is no other part of an organisation where the numbers are so against us – we spend significant time, effort and resources to conduct our safety and risk management activity, but gauging success amounts all too often amounts to seeing that *less* of something happens.

Whether we consider it from a moral, scientific or legal perspective, workplace injuries are essentially outcomes representing failures of the organisational systems designed to prevent them from occurring. In other chapters of this book we look deeper at why these failures occur, but for now let's think a little more about what this really means in terms of measurement. Once we start using Lagging Indicators, such as Accident Frequency Rates, we quickly find ourselves with the dilemma of setting 'improvement targets'. When we set out to reduce our injuries by X per cent this year against last year, we're essentially saying "let's aim to fail less this year!"

Where else in the organisation would it be acceptable to explicitly state that you expect to fail, but will try hard to do it less times than you've done in the past? Would the Quality Department intend to disappoint 10 per cent fewer customers by delivering poor quality products? Would Maintenance plan for 50 per cent fewer unexpected breakdowns? Would Logistics celebrate that they only late-delivered 20 per cent of orders compared to 30 per cent late in the last quarter? Measuring and relying on under-performance doesn't seem like a robust strategy for success. 'Working harder to fail less' doesn't sound like an appropriate boardroom maxim designed to inspire confidence and encourage customer orders, does it?

Scoring own goals

Let's look for a moment beyond the workplace. Manchester United is one of the most consistently successful football (soccer) teams in the world, having won more than 20 league titles, 11 FA cups, five European tournaments and a FIFA World Cup. Under the guidance of manager Sir Alex Ferguson, the club has risen to become one of the richest and most talked-about in the world.

Even if you're not a major football fan, let your mind imagine that it's the afternoon of a big match. Hear the crowd getting excited. Watch Sir Alex as he strides into the players' dressing room for a pre-match pep talk with the team.

As he opens his mouth, to speak, can you imagine him saying "Right lads, just don't lose this game, and all the others this season, and we'll win the league?"

Surely not. When the focus is on 'not losing' the only tactic is defence. Now imagine all those footballers standing in front of the goal, protecting it so that the opposition's ball can't get in. It would be an unusual match, wouldn't it? And one which spectators – and, arguably the players – would soon get bored with.

In pursuit of nothing

In the past decade or so, it's become fashionable, perhaps in an attempt to 'brighten up' the world of workplace safety, to advertise a mission to strive for 'zero accidents' (or 'zero injuries', or 'zero incidents'). On the face of it, this appears commendable, but there are three fundamental issues with such a target.

The first is definition. What do we really mean when we say 'zero' accidents? Do we mean zero 'Lost Time Accidents'? Zero 'Reportable' events? Zero first aid incidents?

Or do we mean 'Zero *Everything*'?

No injury, no matter how small or insignificant, no scratches of fingers, no paper cuts or coffee spills. No property damage, no dropped pallets or forklift trucks scuffing their forks against a kerb? No 'could've been worse' events, such as a slight jarred ankle as we descend a staircase, but catch the handrail to steady ourselves?

Of course, each variation of the definition is skewed. If we say 'Zero Lost Time Accidents' then conversely we must mean that minor accidents where no time has been lost are acceptable. And then of course there is the challenge of duration. Over what period must we achieve zero? From now until then end of the month? The year? Forever?

Second, if we could agree a definition of zero, and then we reached that target, what next? How would we measure continuous improvement moving forward? Would maintaining zero prove that we are safe? On 20 April 2010, senior management from BP and oil rig owner-operator Transocean arrived on a platform in the Gulf of Mexico to celebrate seven years of injury-free production. Just hours after their arrival the Deepwater Horizon oil drilling rig exploded, claiming 11 lives and injuring 20 more in what would become one of the biggest environmental and safety disasters in the world. Was this rig really *safe*? Whilst the Lagging Indicators played out one story, the media revealed another over the subsequent weeks and months, highlighting failings in safety culture, behaviours, leadership and management. Did the rig observe just *the notes*? Was it deaf to the rest of the music?

The third issue with a 'zero' focus is that an accident is essentially a *failure* within the organisational safety system. 'Zero Accident' approaches essentially present the formula that 'Safety = Zero Failures'. It's here that a very subtle, but very dangerous transformation occurs. As human beings, we learn through experience – both positive and in failure. When an individual worker carries out her tasks in a way that does not result in an accident, she may conclude that she has worked in safety. As each day passes, and the worker has not yet suffered an injury, the belief that she is working safely is reinforced and the formula now turns to 'Safety = Anything I do that doesn't cause me injury'. When this belief becomes normalized within a group environment complacency sets fast and risks inherent to the task or process are not noticed. Then, when an accident does occur, it will be met with disbelief and cries of 'but she's never had an accident before'.

Lagging Indicator metrics have an important role to play, but we must bear in mind that they are not the *goal*. This approach will only get us so far, and at some point, a plateau will be reached. Albert Einstein's words 'if you keep doing the same thing, you'll keep getting the same results' resonate well at this point. But beyond this, historical data such as Lagging Indicators encourage a natural reliance on 'what we've done before' and drive the organisation towards a reactive way of working, fostering a fire-fighting culture and reactive approach to safety. Now this may be fine for some, but if you're reading this book, there's a fair chance that you aspire to something more than just maintaining 'good' or average performance. Reliance on a downward trend in the number of accidents you sustain will not necessarily mean that you will continue to head in that direction, and Lagging Indicators offer little guidance on how to remain on the track to continuous improvement.

Systematic reduction in accidents may demonstrate good progress in the right direction, and 'zero accidents' sounds like a laudable aspiration, but these are not appropriate *goal*s for a forward-thinking organisation. After all, how do we really know if our reductions are the direct result of hard work, dedicated effort, luck, voodoo or divine intervention?

We'll look at the concept of 'zero accidents' again elsewhere in this book, but for now, in this chapter, let's think about 'safety excellence' as the target or goal, where through deliberate positive actions a safe workplace is achieved, and, when we achieve this goal, our *outputs* might include a reduction in accident rates, days lost, sickness absence etc.

Leading Indicators

Reductions in accident and injury rates, whilst clearly very important, do not in themselves indicate the presence of safety. This is why we must incorporate leading metrics to measure the 'degree of effort' invested in our process. In OSH management, like in many aspects of life, it is far better to measure and take action *before* an incident occurs, then try to measure what went wrong *after* the event.

The purpose of Leading Indicators is to proactively monitor the efficacy of risk control measures and reflect performance status in a way that builds confidence in a process or activity, highlighting where focus or effort is required to bring things back on track. Leading Indicators can be utilised to measure the 'inputs' that are made as part of the overall safety management system and in this way rather than counting the number of negative events or 'outputs' (such as accidents or injuries) that have occurred, Leading Indicators reflect positive actions that are

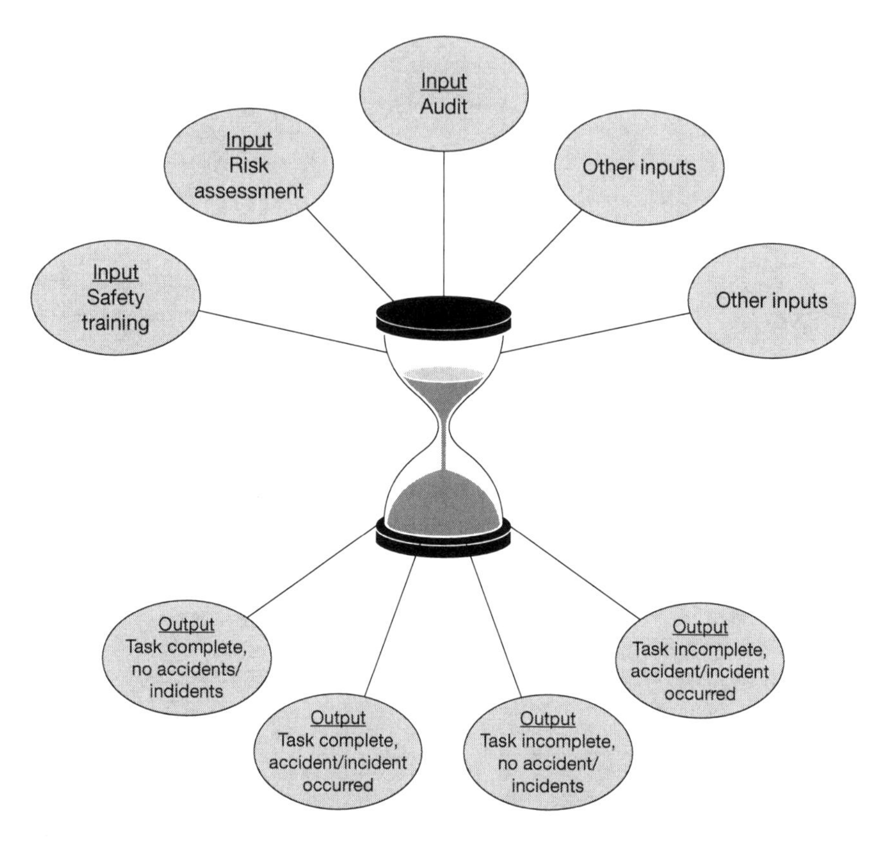

Figure 15.1 Input–Process–Output

taken with the event of *preventing the negative events occurring*. Like most things in life, the more effort or 'input' one makes, the higher the likelihood of achieving ones' goals. Recent research has shown that there is a correlation between amount of management time spent on safety activities and the frequency rate of serious accidents.

In this way, and when used in conjunction with appropriate Lagging Indicators, Leading Indicators can help provide a fuller picture and play an important role in building and sustaining a positive culture towards safety. Of course, as the diagram below shows, there may be many 'inputs' to our approach to managing workplace OSH, so selecting the right inputs upon which to set performance indicators is crucial.

In order to add value, Leading Indicators should be:

- objective;
- clearly defined and understood;
- relatively easy to measure;

- sufficiently demanding;
- specific to activity, task or work area;
- relevant to the organisation, department or work area being measured;
- under the control of the department or work area being measured.

They should also:

- reflect each of the critical or key safety risks identified;
- hold a strong association between the inputs being measured by the Leading Indicators and the desired lagging outputs (e.g. that relevant safety training actually does hold the potential to reduce injuries);
- offer immediate, real-time, reliable indication of performance;
- provide information that accurately guides where improvement is required;
- support long term development of OSH performance;
- not be utilised solely as a vehicle of 'good news' for the organisation.

It's often thought that selecting Leading Indicators is much more difficult that choosing Lagging Indicators, but it doesn't have to be that way.

We all use Leading Indicators in our daily lives – whether gauges on our car, speedometers on our bicycles, traffic lights at pedestrian crossings and more. In the workplace, consider what actions are being taken to *create* safety (the 'inputs'). Think broadly and include behaviours, climate, culture, the work environment, responsibility and accountability. Now consider what is gained in return for these actions. Careful analysis of the activities your organisation is currently engaged in will quickly help identify those activities that add value to your safety endeavours and those which don't. From this position, you will be able to see which actions you then need to measure and set indicators for.

Table 15.1 provides examples of common Leading Indicators.

Health indicators

For the majority of this chapter we have discussed OSH data broadly by referring to safety metrics, however there are many options for measuring the performance in health terms. A common metric is related to sickness absence. In theory, sickness absence data may be easier to collect than accident data, in the sense that an employee absence from the workplace is relatively obvious, at least to their immediate supervisor, and is often recorded within a Human Resources system. There may be variations in absence rates between industries, for example, studies have revealed high levels of absenteeism in call centre environments, and also within organisations, perhaps due to employee satisfaction, sick-pay arrangements, local cultures and perhaps even shift patterns. If you utilize sickness absence rates as metrics for looking at the health of your workforce, remember that just like injury rates, measuring absence is reactive, so therefore a lagging indicator. Be aware too that achieving low levels of absence does not necessarily mean that your workers are 'healthy' (in the same way that achieving zero accidents

Table 15.1 Examples of Leading Performance Indicators – safety

- accident investigation closure time
- accident investigation action completed (%)
- % supervisor/manager time spent on OSH activities
- near misses reported per employee
- ratio near misses to accidents
- number of completed audit/inspection/review activities versus the number scheduled
- OSH audit actions identified
- OSH audit actions completed (%)
- OSH audit actions completed in prescribed timeframe (%)
- % pre-start user checks completed against required
- % statutory inspections completed on time per schedule
- # statutory inspections with X faults requiring remediation
- % training/refreshers completed on time
- % risk assessments/SSOW/SOP reviewed against schedule
- % inspection and testing of key risk controls performed against schedule
- % of inspections where no defects are identified
- # management OSH tours/walkarounds completed
- % management OSH tours/walkarounds against target
- # hours OSH training per employee/department
- % toolbox talks delivered on time
- % of workforce receiving toolbox talks
- # behavioural observations completed per manager/department
- # safe behaviours identified
- ratio of unsafe/safe behaviours identified
- % management time spent on safety (research shows strong correlation between this and performance)
- % management and supervisors job descriptions that include specific health and safety responsibilities
- perception of management commitment to safety
- number of risk assessments reviewed and revised
- number of safety improvement suggestions
- % of identified gaps in training competency resolved
- % of work permits confirmed as completed satisfactorily

Table 15.2 Examples of Leading Performance Indicators – health

- the creation of a worker health plan developed to meet identified workplace health risks
- % of workers who have completed a medical review as required by their job description or role profile
- % of workers who have completed occupational health surveillance against the number scheduled
- % of specific task health risk assessments completed versus planned
- % of specific regulatory risk assessments/reassessments conducted according to statutory frequencies
- % of First Aid/Emergency Response training/refresher training completed against plan

doesn't guarantee we are 'safe') – you may need to consider other performance measurement indicators to give you the full picture.

Just like with the safety examples given above, there can be many other health indicators, related to, for example, access to first aid/treatment/on-site medical or health advice; rehabilitation times; return to-work programmes and their effectiveness. Just as for the safety indicators, setting appropriate health indicators will depend upon the requirements within a given organisation. Table 15.2 provides some suggestions of indicators that could be used.

It is important to bear in mind that when measuring health performance that the latency period between exposure to health risks and the onset of symptoms or disease may be long, so a blend of indicators that look at short, mid and long-term health issues may be prudent. Furthermore, some metrics may relate to the assessment of individuals. Care must be taken to ensure that personal data is always handled sensitively and data presented assures anonymity and is in line with medical confidentiality rules.

Process indicators

Measuring only safety or health data may not provide the full and balanced picture required and may leave the organisation open to low-frequency, high-impact events. On 23 March 2005, at the BP refinery plant in Texas City, United States, an explosion killed 15 people and seriously injured a further 170. Prior to this event, many of the standard OSH performance indicators had appeared positive. This incident highlights that different indicators are required to ensure that a balanced, holistic view is taken that embraces not only the common activities and foreseeable events, but also those that could lead to more serious consequences and major failures. Process Safety Indicators are more complex than other types of indicators discussed in this chapter, and require careful thought on a plant or organisation-specific level. Detailed analysis, consideration and planning of 'worst-case scenarios' in order to identify and manage key risks is vital, before reviewing the control systems in place and selecting relevant metrics that provide accurate reflection of the efficacy of the control systems.

Other types of performance measurement

Auditing

Commonplace in financial areas to gauge performance, and relied upon as robust method of verification. So, should it be in OSH too. But caution, OSH audits tend to focus on management systems, and often forget the softer side of safety, the attitudinal, behavioural and cultural aspects. This may mean that one of the most significant contributors to workplace accidents, human factors, is overlooked. Attitudes to audit findings though can vary. If adverse findings are viewed as criticism or failure which must be suppressed then quickly a negative culture will develop where the audit is tolerated but not embraced, or worse, findings ignored and the process rendered useless.

Measuring attitudes and behaviours

As an organisation's culture matures, we proceed through the phases of improving our plant and equipment, then our policies and procedures and then, when we reach the next performance plateau, we eventually turn our minds to the humanistic aspects of workplace safety. Unlike numerical data, the measurement of human elements related to performance can prove more challenging.

Behavioural observations

The advent of behavioural observations has become a modern phenomenon in OSH with many large organisations following this approach and observation programmes fast-becoming a cornerstone as many workplaces seek to build or further improve their culture towards safety.

Typically, such programmes revolve around trained observers watching how workers perform their tasks, and then a short dialogue commences to discuss observations relative to hazards in the work environment, the safety of the employee conducting the task (or their workmates) or other aspects of risk management.

When the observation programme is launched in a workplace, it's easy to get carried away with the idea that it's a panacea for safety. Small armies of observers are dispatched to the shop-floor replete with checklist, notecards and pencils and targets to complete ten observations this week. Somewhere between lurking in the shadows and brashly stepping up to a worker's shoulder, the observer begins his duty. It's at this point where things can go disastrously wrong.

REFLECTION POINT: LOOKING GOOD

Imagine for a moment that you're about to head out on a date with someone new, or perhaps to a party or business function where you won't know the majority of people there. What do you do?

Perhaps on the way there, you're pretty much yourself, thinking about things in the usual way, walking in your usual way, dressed in your usual way. But as you approach the date or function location what happens? You become aware that you will soon be in the spotlight, so you smarten yourself up a little, stand a little straighter, polish your shoes.

When the conversation begins you may try to appear humorous, well-versed, up to date with current affairs . . . we all want to look good in front of our peers, and can all admit to 'putting on a good front' when required, right?

And so, it is in the workplace. After all, it's human nature to show that we can do the right thing, isn't it? And especially so when, as you are busy doing your usual routine – pressing the button, oiling the cross-arm, striking the mould, repairing a breakdown or sending the product down the line – you become aware of someone watching you. *And taking notes!* We'll talk more about the challenges (and value) of behavioural observations in other chapters, but for now, with regards to performance measurement let's just consider this a 'health warning' to remind us to be clear about what we are really trying to measure. Regular observation and feedback on safety to workers is crucial to raise awareness,

CASE STUDY: HIDE AND SEEK

Three weeks after a global heavy engineering firm implemented a behavioural safety programme across their manufacturing sites, the complaints began to trickle back to management. The workers did not appreciate the one-way process of being formally 'observed' as they worked. Management responded by explaining that the programme was designed to make the workplace safer and advised that graphs and reports would be placed on noticeboards in each work area every week, so that employees could see the results of the observations.

Over the next few weeks some of the reports and graphs mysteriously disappeared shortly after they had been pinned to the boards. In the midst of a busy Production Department, this was normal though, thought the managers. Perhaps someone has borrowed the documents to discuss at a team talk . . .

After three months of the reports continuing to vanish, management met with the behavioural observers to ask for their thoughts. It transpired that several of the observers had been told by colleagues that the process of observing them in silence didn't seem fair. The lack of immediate feedback felt uncomfortable. When several of the front-line supervisors confirmed the discomfort amongst the workers, the penny finally dropped . . .

A team comprised of observers, workers and supervisors was set up to discuss the observation process and agree how feedback would be shared moving forward. Amongst the changes were new ideas to provide immediate feedback following observations, and also to discuss the general headline themes arising from the observation programme at employees' monthly appraisals.

Within two months of the new feedback process being installed supervisors reported that their team members seemed much more engaged at their monthly one-to-ones. The discussion of safety behaviours had even encouraged some to begin to offer their own new ideas on how to respond to the observation results, to make the workplace safer.

drive engagement and maintain focus, but care must be taken with the presentation of data gained. Get the balance right and you may also have the effect of generating some positive interdepartmental competition. Get things wrong and you may wind up with a deterioration of workforce trust and a bunch of numbers that don't have any relevancy.

Attitude, climate and culture surveys

"Statistics are no substitute for judgment." US Congressman and Senator Henry Clay knew a thing or two about using numbers to make a point. But he was wise enough to understand that on their own, they can mislead and misrepresent too. In fact, statistics can allow us to present data in any way we wish, in order to convince our audience that the 'numbers stack up'. So, in order to test whether your numbers are in balance, qualitative data, or opinions, from your workforce may just offer that clearer level of judgement.

In many organisations, 360-degree feedback is common in leadership assessment processes, and listening to the 'Voice of the Customer' has become fashionable in recent years. Safety attitude, climate or culture surveys can help to provide a full 360 of the workplace and allow the voice of the 'internal customer' to be heard.

Utilised to discover the perceptions and beliefs of employees from the shop-floor and from within hierarchical management structures with regard to OSH there is now a broad range of 'off-the-shelf' survey tools that can be used in proprietary format, or tailored to suit the needs of the organisation. Surveys are typically conducted at department, site or company level, either through an online computerised programme, or for those without access to a computer, they can be completed in paper format and fed into the survey software for analysis.

Such surveys can be particularly useful to verify whether attitudes, behaviours and predominant cultures are aligned and in support of the organisational vision for safety success. It may be tempting to dismiss inputs in such survey initiatives as complaints, ignorance, disloyalty, dysfunctions or even carefully crafted posturing from disgruntled or disruptive individuals, but if approached objectively, well-thought out and properly administered they can really help to identify divergence between the expectations and realities of the workplace.

It's important to note that in addition to the survey exercise, it is necessary to qualify whether the attitudes indicated by the survey responses are also observed in the behaviours of employees. If not, or significant variance is found between the reported prevailing attitude and the majority behaviour, then attitude measures as a predictor of OSH performance may be weak.

Where properly constructed, safety survey tools can be an effective conduit to allow the 'voice of the workers' to be heard on matters of safety and risk, provide valid and reliable indicators of safety performance and identify areas for further development, beyond the insight that more traditional quantitative data-driven performance indicators may provide.

Output from such surveys can help to:

- raise general awareness of OSH;
- create a baseline or benchmark for future progress measurement and comparison;
- assess readiness or 'set the scene' for forthcoming OSH initiatives, campaigns or culture programmes;
- understand prevailing attitudes, beliefs and values within the organisation;
- highlight specific areas for action.

Benchmarking

Benchmarking is the qualitative measurement of an organisation's policies, procedures, programmes, strategies, tools etc. with either standard measurements – such as industry data – or with similar peers. Benchmarking can prove useful to:

- understand how other organisations achieve high levels of OSH performance;
- identify what and where further improvements are necessary.

Information gained from benchmarking activities can help improve OSH performance and enhance the reputation of the organisation.

Once an organisation can understand its performance relative to that of others in the field, the desire and motivation to learn from those considered to be leading the field may be stimulated. Beyond emulating others' success, benchmarking can help organisations recognise strengths and weaknesses and act on lessons learned enabling them to build and drive sustainable improvement in order to:

- assess and catalyse safety culture;
- rejuvenate an existing safety approach or management system;
- reduce the incidence of accidents and ill-health;
- improve OSH performance and Return On Investment;
- ensure regulatory compliance.

Beyond these immediate benefits, benchmarking may also help develop and strengthen relationships with suppliers, contractors, customers and peer organisations within and beyond your industrial sector.

As with many comparison exercises, it is common to desire to benchmark performance with those considered 'Best in Class'. However, the challenge here is to ensure that similar measures are compared. The old adage 'comparing apples with pears' is relevant here. Further challenges may be that some managers may not believe the excellent performance data received from their peers.

Many industry sectors operate benchmarking schemes such as *Responsible Care* in the chemical industry, and there are others which are open to all sectors, such as the *Corporate Health and Safety Performance Index (CHaSPI)* in the United Kingdom. Benchmarking schemes are not limited to OSH, and indeed many of the corporate sustainability reporting schemes in operation today have their roots running back to earlier environmental reporting programmes in the 1980s. Since then, many of the schemes have morphed into, or encouraged development of, more broad range benchmarking programmes to encompass sustainability and Corporate Social Responsibility almost as if to echo this trend in organisational annual report documentation. Several schemes operate globally to look at the sustainability performance of organisations. The *FTSE4Good Index* has been designed specifically to objectively measure the performance of organisations that meet globally recognised corporate standards, whilst the *Global Reporting Initiative* ('GRI') enables organisations across the world to assess their sustainability performance and disclose their results in a format similar to that of financial reporting. Whilst these latter schemes are clearly more formal and include high levels of analysis and scrutiny, essentially leading to the publication of your performance, there may be value to be gained by openly disclosing your data and profiling good performance in competitive landscapes. Indeed, the value inherent in utilising validated and respected data from such global schemes is that both employees and management will more likely accept the data as genuine as opposed to being manufactured to drive home a point. Reviewing performance against a peer group may also serve to drive a shift in the corporate mind-set, from complacency to a commitment to change, as leaders become aware of how the organisation compares to others in the industry, sector or geographic region. It is important to bear in mind however, that as many of these schemes contain performance *improvement* criteria as a significant factor when calculating their indices, those organisations already in positions of safety excellence – or indeed finding themselves on performance plateau – may find themselves disadvantaged to some degree. Accordingly, participation in such schemes must be carefully considered in order to balance the effort required to meet reporting requirements with the scrutiny and publicity that follows.

International benchmarking can present challenges. Data from the European Union, The International Labour Organization, the Organisation for Economic Co-operation and Development and the United States Occupational Safety & Health Administration confirm that a lack of standardisation when reporting and calculating accident rates frequently presents difficulties in attaining solid and reliable data accuracy. Variations run from differing classification of 'injuries', application of procedures and concepts, coverage of employed and self-employed persons, inclusion of road traffic events, to perhaps the biggest challenge for those persons keen to engage in benchmarking or comparative analysis with international peers or industries – calculation of accident rates. As previously mentioned in this chapter, the selection and agreement on the definition of appropriate metrics and calculation terms is crucial.

Table 15.3

Type of benchmarking	*Description*	*Purpose*	*Value*
Internal benchmarking	Within one organisation, assessment of departmental or divisional units	Identification of existing best practice	Easy and quick to implement, however may not reveal true Best in Class performance or real innovation
Functional Benchmarking	Benchmarking of specific criteria with the same functional area (e.g. OSH) of other organisations	Identification of 'Best in Class' levels, innovative approaches, new ideas or perspectives	Cost effective method for generating new approaches. Relationship-building. Good corporate citizenship behaviours.
Performance Benchmarking	Assessment of performance characteristics against agreed criteria using peer organisations, trade associations or third parties	Assess relative levels of performance in specific areas	Identify opportunities for further improvement
Process Benchmarking	Considers specific critical processes and operation, typically through the creation of process mapping	Comparison and analysis of critical processes to identify best practice	Often resulting in short term benefits via improvements to key processes
Strategic Benchmarking	Analysis of long term strategies of peer organisations that have delivered high levels of performance	High level consideration of aspects such as core competencies, product development, service delivery, organisational capability	Useful where realignment or revision of organisational strategies are required. Implementation of findings may take significant time
External benchmarking	Typically, utilising participation of several organisations regarded as 'Best in Class' to assess a range of aspects	Provides opportunities to learn from those at the 'leading edge'	Useful where internal best practices are lacking. May provide a fast track to improvement approaches. May take significant resource to assess information and develop sound recommendations

Six steps to setting performance indicators

First, review your existing suite of metrics and indicators. Do you measure what you value, or value what you measure? If it's the latter, then it's time to strip things back and change the game.

1 **Establish responsibility**
A cross-hierarchical team can add great value at this stage when setting Performance Indicators and might comprise:

- Leadership representatives – leaders need to understand the benefits of managing OSH performance and how this can contribute value to the organisation.
- Technical competency – is crucial to bring understanding on how to measure and calculate.
- Workforce involvement – involve workers in the performance management process to build trust and grow interest and engagement.

2 **Set the scope**

- It's not essential to try to measure *every* activity, initiative or aspect of the Safety Management System. Focus on the key risk control systems and expected outputs.
- Identify the key safety risks you wish to observe and gain confidence in their control.
- Set performance indicators to monitor at various organisational levels: Organisation, Division, Region, Function, Site, Department, Team.
- Ensure indicator definitions are clear, concise and unambiguous.

3 **Review existing controls**

- Consider current controls in place for key risks – are they sufficient? Do they overlap?
- Identify opportunities for revision of Risk Assessments.

4 **Identify critical paths**

- Trace the route of key safety risks through their controls.
- Devise appropriate Leading Indicators for each key stage.
- Set tolerance levels for each indicator to define at which point performance must be flagged to senior management.

5 **Establish data collection and measurement protocols**

- Agree how, where and when to collect data.
- Confirm calculation methods to be used.
- Define format and frequency of performance reports (e.g. Dashboards, Triangles, Scorecards, etc.).

6 **Review data**

- Carry our formal reviews of the data at technical expert level to ensure data remains valid and representative of actual performance.
- Conduct meaningful analysis of data to identify themes and trends.
- Review Leading and Lagging Indicators to gauge their connectivity and relatedness. If a significant difference is noted between them, it may be that the Leading Indicator is set too remote from the key risk control to effectively influence the lagging output.
- Consider how to share data across the organisation so that it meets the needs of each level of stakeholder in terms of governance, awareness, focus, direction, etc.

As we have seen, there are myriad indicators to choose from in order to represent the data. Ultimately, we must choose the measurement indicators that we decide are the most relevant, important and useful to our own organisation. Whilst this chapter has advocated utilising a selection of different metrics to reflect performance, it's important not to build a performance dashboard or scorecard that tries to present data on absolutely *every* safety and health metric possible.

Identifying one or two primary or 'key' performance indicators that everyone can understand, define and explain quickly and easily, supported by a couple of supporting indicators is a more prudent approach. In order to maximise the value of OSH performance indicators, look at how OSH can align with other organisational functions, such as Quality Assurance, Lean Six Sigma or Operations. Explore how these other functions define and measure 'excellence'. Are there any synergies that can be leveraged? For example, what's the Quality Assurance equivalent of a Lost Time Injury? Could it be a Customer Complaint, or a Product Recall? How are such things measured by your peers in QA? What can we learn from our colleagues in other functions? Just remember that in order to be effective and useful, there needs to be a strong link between the indicator, the goal and the actual performance of the organisation.

Strengths and weaknesses

Just as in the football analogy presented earlier in this chapter, organisations will never attain safety excellence when they focus only on what *not* to do (i.e. have accidents) and their metrics present only the negative data. So, selecting the right indicators is crucial.

Look at the way progress is measured across the organisation – typically, you may find metrics that show the 'Number' of actions to be completed, the '% Complete' and the '+/–' against the last period of measurement. This format of measurement can be helpful in safety, but it will not paint the full picture on its own.

All indicators have strengths and weaknesses. Challenge is in selecting the right suite of measures for your organisation and recognizing that these will change as you move forward on your OSH journey. A combination of Leading and Lagging is essential.

Outcome-centred measurement

Not having accidents is not an activity, it's an outcome, so by only measuring outcomes we cannot be sure that our activities are correct. The only way to effectively manage safety and work progressively towards our goal of safety excellence is to plan positive activities and then measure our progress against appropriate targets for these. Allow me to digress. In my free time, I like to scuba dive. I especially like to scuba dive with sharks. Big sharks. Often very big sharks. Now, I know that this may at first appear to be dangerous to you, and indeed, I have frequently heard the comment "Diving with sharks? And you're a safety guy?!"

I don't measure the success of each dive by the number of injuries I sustained. I don't look at my fingers or limbs and count the ones that have been injured. Instead, I plan each dive carefully, assessing the risks and taking into account a broad range of conditions and inputs – weather, depth of water, visibility, temperature, time of day, number of sharks and so on – and then decide on my strategy accordingly.

I never plan my strategy with the aim of 'not losing an arm or leg', but instead plan to achieve a successful dive through focusing on the actions I need to take in order to observe the sharks safely – in this case, for example, getting my buoyancy levels just right, moving slowly and deliberately so as not to alarm the sharks, setting a limit on the number of sharks that will be too many for me to dive with, planning an escape route should I need it, testing that my equipment is in good order and functioning properly, wearing appropriate protective equipment.

By carefully planning and doing all (and more) of these positive actions I enjoy productive and safe dives with these magnificent creatures, and, as an output or outcome of these positive actions, have the benefit of not losing parts of my body.

Quality control

Like all forms of measurement, it's not just quantity but the quality we have to get right. Metrics are not the goal. The goal is 'safety excellence' which, like safely diving with sharks (perhaps 'shark diving excellence'. . .), will be achieved if the right people take the right actions, in the right way, at the right times. The indicators are there simply to provide confidence that are actions are correct and supporting the attainment of our goals.

Accuracy and integrity of data

There can be cultural or political sensitivities in measuring accident numbers, with some areas not as keen to share the big picture across the organisation. In weaker organisational cultures, particularly those striving for 'zero injuries' and early on the journey to cultural maturity, employees may feel 'encouraged' to get back to work after an accident so as to avoid 'Lost Time' or serious injury status.

Clear definition of accident types can be helpful, together with raising a shared understanding across the entire workforce as to why accurate, transparent accident reporting is a critical component in your pursuit of safety excellence.

A quick and useful way to gauge the accuracy and integrity of accident data is to develop simple accident triangles for the organisation. In 1931, Herbert Heinrich found that in a given population of workers, for every 300 unsafe acts that occur, there are 29 minor injuries and one major injury. Heinrich's work was persuasive, despite being conducted under limiting conditions and including by his own admission a fair degree of learned conjecture, and it is still widely referenced today.

Some 38 years later, Frank Bird Jr., then Director of Engineering Services for the Insurance Company of North America became interested in testing the legitimacy of Heinrich's theory to discover whether the ratio Heinrich proposed could be applied to any industrial sector.

Bird analysed over 1.7 million accident reports from almost 300 organisations across 21 different industrial sectors. During the exposure period analysed, 1,750,000 employees worked over 3 billion hours. A substantial set of data by any consideration. Bird's study roughly confirmed Heinrich's rations and added weight to the importance of the Accident Triangle tool. Bird went further by adding a fourth tier to his triangle, showing the existence of approximately 600 non-injury events (what we might now refer to as 'Near Misses') for every reported serious injury.

Further studies by the British workplace safety regulator, the Health & Safety Executive have continued to confirm the relationship between each tier of the triangles. This offers us a general indicator of the *likely ratio* that we should expect to see in our own accident data.

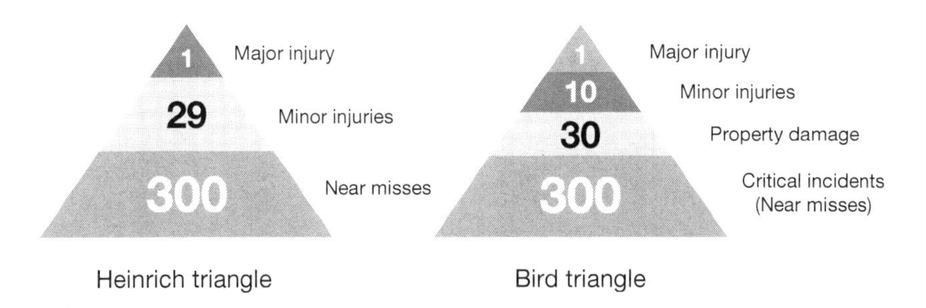

Figure 15.2 Heinrich triangle (l), Bird triangle (r)

Take a piece of paper and create an accident triangle for your own organisation's data and consider how it compares to the research. Get creative and try to draw lines for each side and the base of the triangle that match your data. So, for example, if you have one major injury for every five minor injuries and every seven Near Misses you might end up with a very tall, thin triangle shape. If you had 100 minor injuries and 1,000 Near Misses for every major injury, your triangle would be much shorter and broader.

In accordance with the theories of Heinrich and Bird, what we are hoping for is a general triangle shape: narrow at the top, becoming broader as it moves towards the bottom, with a larger volume of Near Miss or non-injury events at the base. It's not uncommon for some organisations to find themselves with a triangle that ends up looking more like a diamond (one major injury, larger number of minors, very few near misses). If this is the case for your data, and you find yourself with wildly different ratios to Heinrich and Bird, then it may suggest one of the following:

- Your organisation is totally unique. Your people, your plant, processes and events are unlike anything else anywhere in the world.
- Near Misses are not being recognised or reported.
- Serious injuries are being 'downgraded' or under-reported as minor injuries.

Of course, a 'diamond-shaped triangle' may also not mean any of the above, it could point to something entirely different. But this exercise is not a scientific experiment, it's simply an idea to get us thinking about how our data looks in the grander scheme of things. By comparing our own data to a ratio that has survived the test of time we can see whether we are broadly in the right ballpark in terms of accident reporting. If our ratio looks fairly similar to Bird's or Heinrich's, then we gain some confidence in knowing that we may likely have a reasonable level of data accuracy and integrity of reporting.

Before we leave the idea of accident triangles, just a note to point out that both Bird and Heinrich worked for insurance companies. Why is this important to us at this stage? Well, perhaps their studies were driven by an objective of searching for predictability of accidents, which – if identified – could help in the setting of insurance premiums! It's here that much controversy has arisen. In recent years the accident triangle has been called into question for its utility, with critics arguing that it is not a helpful tool to predict the frequency of accidents.

To be clear, I'm not arguing that the triangle offers us a crystal ball for safety, but instead suggesting that it can be helpful as a high-level tool or rule-of-thumb. The triangle provides a simple visual way of sharing data within (and beyond) the organisation – and can be effectively used by several departments, sites or divisions for broad internal benchmarking.

Be aware though, that over the years, the validity of the ratios claimed by Heinrich, Bird et al. have all been criticized. Regardless of any numerical ratio, what is important here in these triangle models is the *principle* at work –

that there is a fine line relationship between each tier of severity. Generally, it is accepted that working to reduce the risk of accidents *at each tier* of the triangle is a worthwhile and impactful activity.

Cause and effect

Once we have created accident triangles, it may be tempting to quickly focus our attention on resolving all Near Miss events in order to reduce these opportunities for harm to be caused and lead to an injury. Whilst focusing on the base of the triangle is logical as it provides us with a broader basis for more effective control of total accident losses, we must remain mindful that there can be a stark difference between the types of situations or events that result in temporary injury and the types that lead to permanent partial or total disability. The causes of these events and outcomes are different.

Systematic analysis to understand the root causes of accidents, ill health and non-injury event trends in your workplace is an essential first step. From here, apply the 80/20 rule to identify where to focus efforts.

Over-measurement

Take care not to fall into the trap of over-measuring, of creating metrics for everything you can think of. Consider your car, motorbike or other vehicle you may operate. How many key performance indicators does it present you with every time you are driving? Three? Four? It's likely that you will have an indicator of your speed, one for the engine temperature, and another for the fuel supply.

You may also have a system that will alert you to a fundamental failure, such as to your vehicle's brakes, whilst you are operating the vehicle. With these basic indicators, you can manage and adjust your inputs, in order to get the appropriate outputs. Other indicators may be present to help you *optimize* the operation – for example an RPM counter, seatbelt reminder beeper, reversing cameras, speed limit warning alarms and so on. However, with the key indicators you should be able to correctly judge the performance of your vehicle and its operation in your hands.

Accountability or blame?

As we have seen in other chapters of this book, fear can be a strong motivator for action (or inaction), and when it comes to measuring performance this is no exception. Go back to my earlier question and let's move a little further: is safety performance measurement principally used in your organisation to hold people accountable for what they did well in OSH – or what they did wrong?

Consider a small child learning to walk, or talk. How do we communicate progress to the child? Do we scold the child each time she mispronounces a word, each time she loses her balance and falls over? Or do we encourage the positive actions like taking well-balanced consecutive steps, or when the words that leave her mouth are well formed and make sense? In the same way, if we only hold people accountable in the workplace for the negative actions – having accidents – then what should we expect in return? Instead of defining progress in OSH as an absence of injuries and sickness absence, redefine it through the positive actions taken to ensure a safe workplace, and hold people accountable for these efforts.

Reward

When results are rewarded – whether for achieving 'fewer accidents' or 'more safety observations' – the behaviours that led to those results are actively recognised and reinforced. But unless you know specifically which behaviours *caused* a particular result, how do you know you are reinforcing the right ones?

Studies show that whilst the provision of rewards can lead to adjustment of short-term behaviours, when the incentive is taken away, behaviours can quickly revert to those behaviours prevalent prior to the implementation of the reward process.

Setting bonus or incentive schemes that hinge on reductions in accident rates may actually be detrimental to performance in the longer term, with accident data being effectively 'hidden' to avoid losing the reward. As a result, the valuable opportunities to learn from accidents and incidents in the workplace is lost, or worse, individuals suffering injury may be encouraged to prematurely return to work. This type of manipulation will quickly erode workforce trust and can lead to cynicism towards other OSH initiatives.

Reward schemes that are based on inputs and Leading Indicators are likely to be successful as they more accurately reflect activities that under the control of the workgroup. Such schemes can assist with galvanising commitment, ownership and engagement in OSH activities and ensure that focus and attention is given to the areas that are considered by the organisation to be the most important.

Before you implement any kind of reward programme linked to safety and risk management, think carefully about the behaviours it's likely to foster. The most effective reward schemes are those that have a clear link between the activity, the performance metric, the overall goal and the actual performance of the organisation.

The importance of feedback

The nineteenth-century lawyer, mathematician and radical freethinker Karl Pearson was fanatical about data. In addition to shaping many of the ideas that were later brought into common understanding by Albert Einstein, Pearson excelled in identifying correlations between data inputs and outputs.

Pearson's 'Correlation Coefficient', or more simply 'Pearson's Law', is one of his most revered – and perhaps one of his most easily comprehensible – concepts. It states that:

> Where performance is **measured**, performance improves.
> Where performance is **measured and reported back**,
> the rate of improvement accelerates.

Throughout his studies, Pearson repeatedly found and underlined the importance of feedback, no matter which construct or application was being measured. Pearson's Law is as relevant today as it was back in the 1880s and serves well to remind us that across the organisation, performance indicators are not the exclusive domain of the boardroom or the OSH department. In order to gain the full benefits of accelerated improvement, think carefully about how you communicate your results with all of your stakeholders, whoever they may be.

Heinrich or Bird's Triangle is a common and often useful way of presenting data to the workforce (and leadership teams) but bear in mind it only tells part of the story – the ratio between different aspects of (principally) lagging performance data. Using a triangle or pyramid device to reflect data can indeed be very useful, but the impact and value may be significantly increased through the addition of brief narrative, or by incorporating the triangle into a broader dashboard or scorecard that can be shared across the business.

THE NAKED TRUTH: PERFORMANCE MEASUREMENT

As Deming (1982) proclaimed: "What gets measured, gets done." A clearly defined safety approach to measuring performance will provide the organisation with a practical means to take the right action.

The purpose of measurement is *not* data collection. It's not to simply have some numbers to show the board. The purpose of measurement is to *understand* where we are now, and to *motivate and inspire people* to *focus the effort* in a particular direction to achieve a particular result.

We would not claim that we are 'healthy' simply because we don't have a disease. Likewise, the idea of an organisation being 'safe' simply because there are no accidents is equally crazy. No single indicator can provide a full, clear and accurate perspective of OSH performance across an entire organisation. A mix or 'balanced scorecard' approach with both leading and lagging indicators is essential.

As an organisation's cultural maturity towards safety develops, performance indicators will likely change, and the use of leading (rather than lagging) indicators will grow more important.

Using a well-thought-out blend of leading and lagging performance indicators, that contain both humanistic and systemic measures and view performance from a range of perspectives will help raise awareness and provide a full and clear picture of performance that allows focused action to be taken to prevent deterioration or failure proactively and *in advance* of adverse events. Utilising qualitative measurement tools such as employee opinion surveys or questionnaires, or safety climate/culture assessments will help you to take a balanced view.

GET NAKED: PERFORMANCE MEASUREMENT

1. How do you currently measure safety performance in your organisation? Do you rely on lagging indicators or have a blend that includes solid leading indicators too? What does this approach tell you about your safety culture?
2. Do your performance indicators reflect each of the key safety risks you have identified? To be effective, there needs to be a strong correlation between the indicator, the goal and actual performance. Be clear to measure what you value, rather than valuing what you measure!
3. If you conducted a safety attitude or culture survey right now in your business, what do you think the top three findings might be? How are these reflected by your current OSH performance?
4. What does 'safety excellence' look like in your industry sector? How do you compare to the 'best in class'?
5. What do you do with the data obtained through OSH performance measurement? How do you share this across the workforce? Would employees be able to accurately articulate safety performance in ways other than relying on accident or injury frequency rates?

16 Health, work and wellbeing

When talking about occupational health and safety, the subject of employee wellbeing is often absent from discussion.

In the spirit of stripping back, we'll explore the roots of the term well-being and follow them as they weave a path towards effective workplace interventions.

Drawing broadly on a range of academic research and scientific studies we will consider how modern organisations might seek to enhance the power of their people with Bach's Brandenburg Concerto Number 3, *a brisk walk and a smile.*

Finally, in this chapter, we'll discover the future of health and wellbeing by revealing four megatrends and learn how other organisations have tackled these issues successfully.

Former British Prime Minister Winston Churchill's words are as relevant now as they were 75 years ago, when he uttered them on 21 March 1943. Churchill was elaborating his four-year plan for England and took the opportunity to underline the importance of health, strength, fitness and vitality to a nation caught in a period of significant change.

Today, a distillation of Churchill's maxim frequently reappears in the world of work as 'our people are our greatest asset' adorning everything from reception foyers of corporate offices, vacancy advertisements and Annual Reports. Yet, the concept of employees as a precious resource to organisations is nothing new. In the seventeenth century, the 'Father of Occupational Medicine' Bernardino Ramazzini argued that workers' health should be both protected from risks at work and, wherever possible, 'optimised'.

On 22 July 1946, representatives of 61 states came together at the International Health Conference in New York, USA to sign the World Health Organization's (WHO) formal constitution. On page 100 of the constitution, the country leaders declared: "Health is a state of complete physical, mental and social well-being and not merely the absence of disease or infirmity" (WHO, 1946).

The declaration served to set solid foundations for future dialogue on health, but beyond these important duties, it introduced the concept of wellbeing to the world. This inclusive definition of health has stood the test of time, and over 70 years later it has never been amended, and continues to be utilised in many WHO publications, Non-Governmental Organisation guidebooks and within academia.

The development of strategies for managing the health and wellbeing of the workforce has been positively promoted since the WHO constitution, and in recent years health – and, in particular, wellbeing – have been at the heart of many new initiatives driven by NGOs, governments, regulators and research institutes alike. This focus may be encouraged by issues such as changing population demographics, challenging financial climates, the extension of working lives or perhaps even altruism. But regardless of the catalyst for action, as workplace risks move from the physical to the psychological, an increasing shift towards enhancing wellbeing and optimising health, can be observed on organisational landscapes across the globe.

The growing cost of ill-being

The terms 'mental health', 'psychological contract', 'burnout' and 'work-related stress' have now all entered common parlance, with the latter becoming a particular concern. Why this attention? The simple answer is money. Poor health costs money. With the cost of work-related health and associated productivity losses now standing at around 5 per cent of global Gross Domestic Product (GDP), the economic incentive for businesses has never been greater.

In the more developed countries, mental ill health has fast become one of the biggest causes of work absence. The Health & Safety Executive calculates that the cost of work-related stress to UK society is now somewhere in the region of £4 billion a year, and with over ten million working days lost in the UK in

2016/17 it is no surprise that modern organisations are keen to take action to improve employee health and wellbeing. In the United States of America, every day just over one million workers call in sick complaining of stress, leading to a $200 billion a year cost to employers. But whether considered in the staffroom, boardroom or courtroom, conversations around 'stress' and 'mental health' have time and time again proved awkward, sensitive and fraught with subjectivity. Organisations that decide to take action often find themselves clutching at loose ends, unsure which way to move.

But it's not all in the mind. The United States leads the globe spending $2.7 trillion each year on health care – more than any other country around the world, per capita. Yet, statistically, it regularly ranks in the bottom quartile for health and wellbeing. One in every three Americans is now classified as 'obese' – the term given where excess body fat has accumulated to such volume that it is considered hazardous to health, typically leading to myriad ill-health conditions and reduced life expectancy. A further 31 per cent of the American populace are considered overweight, and at the current rate, by the year 2020 obesity-related diseases will overtake smoking as the biggest killer of American people and account for 20 per cent of all health care costs to the third biggest nation in the world. As we progress through the Knowledge Era we find ourselves slowing down more than ever before. In 1963, almost half of all jobs in developed countries required some element of physical effort. Just 50 years later, in 2013 less than 15 per cent of work roles require such effort, leaving more than 85 per cent of us sedentary at our workstations. As a result, we now burn 140 fewer calories per day than the average worker did back in 1963. In America, only 6 per cent of all jobs now meet the US Government guidelines for physical effort in the workplace. All this inactivity is doing us no good. People with sedentary work roles are more than twice as likely to die from heart disease than those with mild to moderately active jobs. Doing nothing is making us ill.

The World Health Organization advises that 217 million cases of occupational disease are now diagnosed each year. Of these, 1.6 million lead to loss of life. With between 1.9 and 2.3 million deaths each year attributed to occupation, optimising physical and mental health at work is not just a moral imperative, but an organisational one too; to ensure that our workforce contributes fully to corporate efficacy and success, and enhances global competitiveness. Improving health and developing wellbeing to adapt and thrive in a fast-changing and uncertain world benefits both the individual through increased wellbeing and resilience, and the organisation by helping transform risk and turmoil into opportunity and stability.

A new rhetoric?

Although a slow burn since its introduction by the WHO constitution in 1946, the rise of the term wellbeing may indeed provide a welcome diversion from the negative stigma surrounding 'stress', the weight of obesity and the challenge of occupational disease. But as a universal concept, despite the growing interest in 'wellbeing' at work (like much of the psychosocial language), agreement on its definition, and, importantly, the effectiveness of interventions, is rare.

There have been many attempts in recent years to explain just what it is we're talking about when we use the word 'wellbeing':

- 'a composite measure of how good an individual feels at the physical, mental and social level' (Szalai and Andrews, 1981)
- 'the subjective state of being healthy, happy, contented, comfortable, and satisfied with one's life' (Waddell and Burton, 2006)
- 'the combination of feeling good and functioning effectively' (Huppert, 2008)
- 'a summative concept that characterizes the quality of working lives, including occupational safety and health aspects' (Schulte and Vainio, 2010)

Of course, the definitions of wellbeing proliferate, but each seem vague, and without agreement, what should we conclude? Well, without trying to over-complicate things here, we might consider that 'wellbeing' is a self-measured feeling of inner health and happiness that is affected by interactions with others and our environments.

Happy, healthy and here

Whilst it's easy to hypothesise that healthier workers will be happier and more productive, several recent studies do confirm strong links between wellbeing and productivity at individual, organisational and societal levels. Recent findings from Australian research suggests that a healthy worker can be up to three times more productive than colleagues in poorer health (Medibank Private, 2005). In challenging times, businesses may increase productivity and strengthen their competitive edge by enhancing employees' wellbeing and developing their 'psychological capital'.

In a report released in 2013, it was revealed that more than a quarter (27 per cent) of the entire UK workforce have a health problem that has lasted for more than one full year. The British Heart Foundation's research revealed that 12 per cent of these workers felt that their poor health was limiting their ability to do their job. The Health at Work Index report further identified that one in six of these health problems is associated with either heart or circulatory problems, the risks from which can be significantly reduced by simply losing weight, increasing physical activity and making healthy eating choices.

Beyond the immediate financial costs and potential for productivity enhancement, research by Robinson and colleagues reveals that the longer an employee is absent from work due to ill health, the lower the probability of their return (Robinson et al. 1997). After a 6–month-long absence there is a 50 per cent chance that the employee will not return to work – CIPD 2003. At 12 months of absence, the likelihood of returning falls to just 25 per cent, and after two years the chance of the employee returning to work drops to practically zero. Looking at mental health in isolation, one of the biggest health risks facing people in more developed nations, 40 per cent of those individual employees absent with mental health issues for a period greater than five weeks will not return to the workplace.

Even in cases where the employee does return to work, after five weeks of absence a marked reduction in their ability to work may be observed. This may be down to individual coping styles or levels of personal confidence, as research by Pearson (2001) concludes that belief in one's own dysfunction or ability to return to work and continue in the same way as before is a significant factor in influencing the duration of absence. Pearson's study goes on to highlight that the longer the absence, the stronger the belief in one's lack of ability to fit back in to the work environment and regain feelings of effective contribution. In challenging post-recessional times, the prospect of losing skilled labour cannot be understated and employers would be wise to do all they can to expedite the return of employees to the workplace. Time is of the essence. Taking early interventions to support the rehabilitation of a sick worker can result in a significant reduction in medical practitioner visits and reduces both the risk and duration of any subsequent work absences.

The impact of lifestyle

Each year, non-communicable diseases (NCDs) such as cardiovascular conditions, cancers, diabetes and chronic lung disease are responsible for around two-thirds of all the deaths of human beings around the world.

These diseases killed more than all the uprisings, clashes, conflicts, insurgencies, political violence and wars across the globe combined. And many, many of them are strongly influenced by the lifestyle choices *we* make.

The proportion of deaths from each of the 'big four' NCDs – heart disease, stroke, respiratory and lung diseases – has risen significantly since the year 2000. Collectively, deaths from these NCDs have risen more than 60 per cent in the first decade of the twenty-first century. Cardiovascular disease such as heart attack, stroke and hypertension or high blood pressure currently kills an estimated 17 million people each year – and this figure is set to rise to around 24 million by the year 2030.

What's behind each of these killers? Our lifestyle. Despite the best advice, we continue to eat more, drink more, smoke more and exercise less than we know is good for us. No matter which part of the world we live in, heart disease and stroke carry the greatest risk for all of us, and feature in the top five causes of death no matter which income level we fit into. It's a darker picture in many of the most economically developed countries including America, Australia, Great Britain and the European continent where despite – or perhaps because of – our ability to generate higher incomes (which support perceived higher standards of living) the chances of dying from NCDs is greater than in lower income nations. Our modern lifestyles involving longer working hours, sedentary work, frequent travel and love of socialising are harming our wellbeing and literally causing our bodies to fall apart at the seams.

Time for action

We spend about 28,000 days on this planet, and around 10,500 of those – or about 37 per cent of our waking hours – are spent at work, (Smith Institute, 2005) so the workplace presents a fantastic environment to reach individuals with messages on health and wellbeing. But how does an organisation choose where to focus? With data as strongly persuasive as the previous data on NCDs, it's easy to begin our campaigns here, but this presents us with a paradox. By focusing on disease, we have immediately turned the concept of wellbeing on its head. Instead of discussing *well*ness, we begin by talking about *sick*ness . . .

But beyond this criticism, providing a balance, or 'sweet spot' is found between raising awareness of disease causes and symptoms and encouraging the attainment and maintenance of a healthy lifestyle, everyone can benefit.

Everyone? Yes, it's not just the high-flying business leaders that succumb to the killer NCDs. Factory workers, office staff and everyone in between are just as likely to encounter coronary, pulmonary, circulatory and mental health issues. The International Labour Organization (ILO) estimates that each year, around 36 million deaths can be attributed to Non-Communicable Diseases such as these 'big four'. And musculo-skeletal disorders are beginning to carve out a place for themselves as a major cause of work related ill health too, in both sedentary work such as data entry and administration, and also in repetitive activity or jobs that require high levels of physical effort.

In the less economically developed countries there is the potential for significant variation in causes of death between countries. Whilst it can be expected that in these nations a major proportion of deaths will be caused by infectious diseases such as HIV, malaria, tuberculosis, etc., alarming increases in NCDs, especially heart and respiratory diseases can be observed, year on year. It would appear that as the economy increases, so too do the diseases, for example as development continues in China and South East Asia, the desire to mirror the cultures and practices of developed nations through the adoption of 'Western' diets, and exposure to more 'sophisticated' styles of alcohols, grows.

With this context, a good starting point could be to target attention towards reducing the risk of the four main non-communicable diseases, given their prevalence across the globe, and the continuing rising trends. Consider building interventions around established issues like diet, exercise, alcohol, smoking and stress/mental health.

Beyond broad-brush interventions focused on the big four, it's worth bearing in mind that even within economic regions there is the potential for further variance, take for example the stark contrast between the stereotypical deep-fried fat-laden foods of the United Kingdom, versus the healthy 'Mediterranean diet' in countries such as in Italy or Greece where people typically eat more than six portions of fresh fruit and vegetables every day.

Look at the demographics of your workforce and consider interventions aimed specifically at your target audience. In women's health, for example, 90 per cent of ovarian cancers occur in woman aged 45 years plus, 80 per cent of breast cancers are diagnosed in post-menopausal women (typically aged 51 years plus), and one in every two women over the age of 50 will suffer an osteoporosis-related bone fracture.

For interventions aimed at male workers, did you know that prostate cancer is one of the most common cancers in men, affecting one in every eight men in the UK, and black African and black Caribbean men are three times more likely to develop the cancer than white men. Bladder cancers are the fourth most commonly diagnosed cancer in American men. Smoking is responsible for around 50 per cent of all bladder cancers, and smokers are three times more likely to contract the disease than non-smokers.

If you are already engaged in delivering wellbeing interventions in your workplace, have you considered some of the newer, emerging issues such as the global obesity crisis? Forty per cent of adults around the world are thought to be overweight, and up to 15 per cent clinically obese. Over three million deaths each year caused by obesity, contributing to 50 per cent of diabetes deaths, 25 per cent of heart disease deaths, and up to 40 per cent of certain cancer-related deaths.

Interventions to support the ageing workforce are also worth exploring. Issues related to deteriorating hearing and eyesight, and worsening mental health problems from dementia and Alzheimer's disease to simple forgetfulness.

Beyond the moralistic or humanitarian purpose, there are many reasons why organisations may be keen to invest in the health and wellbeing of their employees. Frequent aims for interventions in organisational wellbeing programmes include:

- stress reduction and improved coping skills;
- increased job satisfaction;
- reduction in ill-health related absence;
- increased commitment and motivation;
- improved worker engagement and job satisfaction;
- improved personal effectiveness;
- reduced accidents, health care costs and lowered insurance premiums.

Regardless of choice of intervention, the objective is likely to ensure employees are 'happy, healthy and here' as forward-thinking organisations strive to provide encouragement to facilitate behavioural change, reduce sickness absence, decrease employee turnover (or 'churn'), improve productivity, efficiency and quality, boost morale, build industrial relations and enhance corporate reputation. All of this stacks up to better OSH performance.

Interventions

Many things may be considered as 'interventions', designed to influence the relationship between work and wellbeing – from fruit baskets in the workplace, healthy options in the canteen, smoking cessation support, organised fitness activities or information sessions on health risks and disease – yet, few have been evaluated robustly, and many focus on either reducing stress or increasing satisfaction.

Interventions vary by degrees to which they are preventative (reducing risks) or curative (healing the negative effect of risk) and are typically implemented at either Individual level to increase positive aspects of wellbeing or prevent deterioration; or at Organisational level to improve (or prevent reduction in) performance and effectiveness. Wellbeing interventions can be implemented on three different levels, let's look at these now.

Primary interventions

Primary interventions are typically delivered to a broad audience, aimed at removing negative aspects of the workplace and increasing satisfaction. Exponents of primary interventions typically believe that by reviewing the work environment and task demands, and then systematically reducing those job conditions that cause strain, positive behaviours – such as improved attendance at work – will be increased.

Interventions of this type are typically based on Hackman and Oldham's (1976) Job Characteristic Model – however over the years this model has received very little support. The model assumes that certain job characteristics lead to specific psychological states, which in turn lead to outcomes of motivation, satisfaction and effectiveness. Evidence is mixed as the link between increasing certain job conditions, such as autonomy, has also been shown to result in both job satisfaction and dissatisfaction and perhaps now this once popular approach has reached the end of its useful life. As interest in developing role profiles, employee assessments and development reviews has grown significantly in recent years, we could consider that the Job Characteristics Model has been effectively subsumed into broader human resource best practice.

PRIMARY INTERVENTIONS – SOME COMMON EXAMPLES

Absence Management – clarity on sickness absence can identify absence trends and health and safety risks within the organisation.

Drug and Alcohol Screening – can help an organisation develop relevant policies to reduce the risk of employees working whilst under the influence. Such an approach is very common in environments involving high risk activities such as energy, the use of vehicles, heavy machinery or hazardous substances. Screening may range from regular pre-work validation to ad-hoc randomized sampling, or 'with cause' testing.

Employee Health Surveys – online or paper-based questionnaires can help in profiling the overall health of the organisation's workforce and generate a benchmark on health and wellbeing. Whilst they may generate a mass of data, surveys can quickly identify issues upon which the organisation can drive focus and raise awareness.

More sophisticated surveys may provide individual personalised health and wellbeing reports for participants.

Fitness and exercise opportunities – from the organisation of staff walking or jogging or exercise clubs, to onsite fitness facilities, or incentives to join local gym facilities, opportunities to bring exercise closer to workers abound.

Health and Dental Insurance – private medical insurance is typically available across a broad range of cover options, expediting and typically covering the costs of treatment or emergencies.

Health Promotion – offering information and advice on health and wellbeing across a range of issues from smoking, diet, disease, nutrition, sleep, stress and more can be a useful way to raise awareness of wellbeing in the workplace. These interventions are typically passive in delivery and are often 'self-help' programmes using posters, leaflets, emails and inter/intranet resources.

Secondary interventions

These interventions focus on 'at-risk' groups and are implemented to help people learn about health risks and teach them how to change their behaviour in order to improve wellbeing. These interventions typically take the form of training sessions or self-help materials on topics such as stress, time management, coping skills, relaxation techniques etc.

With many such approaches available as 'off-the-shelf' packages, these generic 'self-help' interventions can appear attractive to employers, though there is little scientific evidence available to support their long-term positive impact on employees' wellbeing or performance. Beyond this, robust evidence for their general effectiveness is also limited, meaning that the happiness and wellbeing gained from investment in such programmes remains solely with the employer's self-satisfaction for being seen to have 'done something' for the workforce.

SECONDARY INTERVENTIONS

Case Management – a dedicated service to manage the return to work process after an employee is absent due to sickness or injury can assist in the assessment of the employee's medical needs, and provide guidance on their abilities with regard to specific work tasks.

Health Surveillance – may typically reflect the obligations set out under health and safety legislation, surveillance ensures that employees are fit to perform their duties. A range of tests to check the efficacy of eyesight, hearing, lung function, are common – however, depending on the work environment may include more hazard specific surveillance including dermatological checks, radiation exposure and more.

On-site Health Appointments, Check-Ups and Reviews – the provision of a doctor, nurse or physiotherapist within the work environment provides an accessible point to seek basic medical and health care and may help to reduce absence from the workplace to attend external service providers.

Stress Management Programmes – usually delivered via online self-guided learning modules to help individuals learn about triggers, symptoms and controls. Managers may benefit from more formalized classroom or specific training to help them identify, manage and prevent stress in the workplace.

Tertiary interventions

Tertiary interventions are aimed at the individual level, usually following some kind of issue, such as trauma, disease, pain or illness, through the provision of counselling or Employee Assistance Programmes offering health advice, legal assistance, etc. Whilst we can be argued that counselling is effective in reducing negative affect, its impact on employee behaviour and effectiveness remains unclear.

TERTIARY INTERVENTIONS

Several of the preceding interventions may operate from Primary, through Secondary and into Tertiary level. **Case Management, Drug and Alcohol Screening** and **On-Site Health Appointments** may all prove beneficial activities at each stage.

Employee Assistance Programmes – provide individual employees with confidential advice and guidance on a range of issues from health care, nutrition, disease, and may include other broader domestic issues. Employee Assistance Programmes are usually delivered through telephone conversations, however some more advanced programmes also offer follow up face-to-face sessions with qualified advisors.

Put on a happy face

Across cultures around the world, people consistently rate subjective wellbeing as the most important thing in their lives – more so than success, income and frequently even ranking above family ties and personal connections (Diener, 2000). Despite what we may at first expect, wellbeing is not directly related to socioeconomic status, level of income, level of education, gender or race. So, the young administrative assistant cycling to work in their old tennis shoes may be as happy as the business leader driving to the office in her top of the range BMW and Christian Louboutin heels.

Interest in improving our wellbeing has grown immensely and in recent years the meteoric rise of self-help, personal improvement and spiritual enlightenment books, apps, online forums and blogs certainly serves to underline this prioritisation and fuel a constant loop of generating interest, providing satisfaction and then generating further interest.

With the surge in popularity and interest, the definition of the term wellbeing has also become easier, with both the personal self-development authors and the social sciences moving to define the previously hard-to-classify term of 'subjective wellbeing' with more discernible clarity. Nowadays, wellbeing is about 'feeling good', or perhaps in one word: 'happiness'.

The heritability of happiness

The pursuit of happiness seems to intrigue psychologists as much as it intrigues media journalists, Hollywood film makers and society at large. Research suggests that we may be born 'happy' – inheriting our cheerfulness and subjective wellbeing from our parents. According to recent studies, we are all born with a level of happiness, known as the hedonic set-point. Fortunately, for most of us, our set-points are usually above zero – i.e. on the happy side of neutral.

In a fascinating ten-year study looking at subjective wellbeing in more than 2,300 individuals, social psychologists David Lykken and Auke Tellegen at the University of Minnesota, USA, found that almost 90 per cent rated themselves to have high levels of subjective wellbeing and long-term happiness. It could well be, then, that as we human beings have evolved, our forebears that were grouchy or miserable fared less well in the 'struggle for survival' and had less luck in the mating game, leading the researchers to suggest that mankind has evolved a bias towards positive wellbeing simply through the process of natural selection.

But even if you were in the minority that didn't inherit the happiness gene, your personality may have already taken over to generate a higher level of wellbeing for you. Several cross-sectional studies have shown that wellbeing is strongly related to the Five-Factor Model (FFM) of personality, especially the factors of Neuroticism, Extraversion and Conscientiousness. As many pre-hire personality profiling exercises utilise the FFM this may offer the potential to screen for 'happier employees' – but caution is needed as, typically, they only offer a perspective at a single point in time.

Whilst subjective wellbeing is not entirely subsumed by personality, the two constructs have been reliably correlated by the research for more than twenty years now. Elements such as coping styles, emotional intelligence, sociability and conflict resolution skills have all been strongly linked to helping us feel happier longer (Lyubomirsky et al., 2005).

REFLECTION POINT: JOHN LENNON AND THE PURSUIT OF HAPPINESS

Looking back on his life, John Lennon recalled that when he was five years old, his mother counselled him that "Happiness is the key to life."

Some years later, whilst at school, the young Beatle was given a written assignment to answer the question 'What do I want to be when I grow up?'

John thought long and hard, and answered the paper simply with one word: "Happy."

Failing the assignment, his teacher suggested he hadn't understood the question.

John suggested that the teacher hadn't understood life.

Satisfaction guaranteed?

In a meta-analysis of 99 cross-sectional and longitudinal studies, psychologists Julia Boehm and Sonia Lyubomirsky find that: "Happy people are more satisfied with their jobs. They perform better on assigned tasks than less happy peers and are more likely to take on extra role tasks such as helping others" (Boehm and Lyubomirsky, 2008).

If, from a commercial perspective, the prospect of gaining discretional work activity from employees is not quite enough of a motivator, the study also concluded that happy people are less likely to exhibit withdrawal behaviours such as absenteeism.

The Institute of People and Performance suggest that happier people are:

- 47 times more productive than their least happy colleagues;
- set themselves higher work goals;
- work with 25 per cent more efficiency and effectiveness;
- contribute a day and a quarter more every week than those unhappy workers; *and*
- embrace task objectives with more optimism and visions of anticipated success.[1]

In the United States of America, Gallup's Employee Engagement Index explores employees' attitudes to work and their wellbeing and reveals that engaged employees feel more emotionally connected to and involved in their work activities. These employees smile and laugh more, experience more enjoyment in life and experience less negative feelings such as worry and anger. By contrast, those who are not emotionally connected to their workplace are less likely to invest discretionary effort, and those who are 'actively disengaged' are not just emotionally disconnected, but also jeopardise the work of those around them, and in fact are less happy and have poorer levels of mental health than those people who are unemployed.

In the first half of 2015, 30 per cent of American workers felt enthusiastically engaged in their work activities, compared to 52 per cent not feeling engaged, and 18 per cent actively disengaged. This emotional disconnection saps personal happiness and wellbeing, and reduces the impact and effectiveness of the worker's contribution. Those workers who feel engaged maintain a positive mood throughout the week and benefit from steady levels of happiness compared to their colleagues, and report feeling as good on weekdays whilst at work, as they do on the weekend.

But that's not all – disengaged employees not only have fewer happy days at work, but also suffer worse weekends than their colleagues. This could be due to negative work-related emotions drifting over into leisure time and family life. It's feasible to expect that such elongation of negative emotion, coupled with the lack of respite typically afforded by the weekend may produce harmful effects on long term health.

Whilst the Gallup indices can't provide us with details on the causal relationships between engagement and wellbeing, they do clearly suggest that those individuals who are actively engaged in their work activities appear to be happier and experience less negative feelings than colleagues who are not engaged or are disengaged.

Increasing engagement seems to improve the subjective wellbeing of employees. In turn, improved wellbeing has been shown by other studies to enhance work performance and play a causal role in the achievement of positive outcomes and goal attainment. (Lyubomirsky et al., 2005). So, by engaging our employees fully in the workplace we can maximise opportunities to improve their health, happiness and personal success – and at the same time boost business productivity.

Table 16.1 Daily experiences: unemployed vs. employed job engagement status

	Unemployed	*Actively disengaged*	*Not engaged*	*Engaged*
% smiled or laughed a lot	80	70	86	93
% experienced enjoyment	81	69	89	95
% experienced worry	49	46	35	21
% experienced anger	20	24	13	9

Source: Gallup Employee Engagement Index and Gallup Healthways Wellbeing Index.

The Happiness Advantage

Having what the positive psychology movement calls the 'Happiness Advantage', or an optimistic, positive mindset has also been shown to help doctors improve the speed and accuracy of diagnosis by almost 20 per cent (Estrada et al., 1997), raise the hit-rate of salesmen by over 37 per cent (Lyubomirsky et al., 2005), increase the productivity and job satisfaction of office workers by more than 30 per cent (Sparr and Sonnentag, 2008; Spector, 2002), boost creativity (Estrada et al., 1997), encourage original thinking, build resilience (Grawitch et al., 2003; Estradaet al., 1994) and enhance flexibility (Estrada et al., 1997).

If you're happy and you know it . . .

In a recent global study of the wellbeing of 18,687 individuals across 24 countries, 78 per cent said they were "happy" in life, and 22 per cent indicated that they were "very happy".

The happiest people on the plant reside in Indonesia, where 51 per cent of those surveyed claimed that they were "very happy". India comes a close second place with 43 per cent of people feeling very happy, compared to 30 per cent of those living in Turkey.

By contrast, in countries perceived to enjoy a more relaxed way of life, such as the United States and Australia, only 28 per cent of those asked considered themselves very happy (Ipsos survey, February 2012).

Why are some people happier than others? Lykken and Tellegen's observations of more than 2,000 people concludes that:

> The transitory variations of wellbeing are largely due to fortune's favors, determined by the great genetic lottery that occurs at conception. We are led to conclude that individual differences in human happiness – how one feels at the moment and how happy one feels on average over time – are primarily a matter of chance
>
> (Lykken and Tellegen, 1996)

Lykken and Tellegen suggest that once our personal level of happiness is assigned to us, it becomes fixed and immune to significant fluctuation. They rather pessimistically offer that attempting to enhance one's happiness is a pointless exercise, and that: "Trying to be happier is as futile as trying to be taller."

Fortunately, for the optimists reading this book, recent research contests this bold assertion. If we look at what causes happiness we see that from a scientific perspective it's when the chemical dopamine floods our system, switching on the learning centres in our brain, generating feelings of being alive and outwardly exhibiting signs of happiness.

That flow of dopamine occurs when we are in the right frame of mind. Whilst it's not quite as easy as instructing those around us to 'get positive,' there are four simple actions that have been proven to have almost guaranteed levels of success in helping us attain that 'happiness advantage':

- **Being thankful** – identifying three things to recognise their value and say 'thank you' for (Emmons and McCulloch, 2003)
- **Exercise** – gentle raising of the heart rate through short duration moderate exercise (Babyak et al., 2000). Ten minutes of activity such as jogging, body weight exercises like press ups or sit ups or even walking at a fast pace are all sufficient
- **Meditation** – cultivating a sense of openness or a 'growth mindset' (Dweck, 2007) through self-guided or instructor-led meditation
- **Random acts of kindness** (Lyubomirsky et al., 2005) – simple expressions of altruism including opening a door for someone, allowing the car ahead to exit the junction or bringing a colleague a coffee all count

Of course, there's a catch. Just engaging in one of these four actions in itself is not sufficient. The scientists behind this research agree that in order to rewire the brain and gain the advantage, we need to stick with the chosen new actions for a period of least 21 days. At this point, the actions turn into habits and start to elevate the Dopamine levels, bring smiles to our faces and happiness to our hearts.

The benefits of happiness don't just end with improved wellbeing. A study conducted by American strategy consultants the Reliability Group in 2010 identified the most common underlying causes of safety accidents. Perhaps surprisingly, *cheerfulness of the workplace*, *levels of stress experienced* and *job satisfaction* all ranked within the top ten.

The sound of music

But what if we don't live in Indonesia, don't happen to have been born with the happy gene or are unable to find 21 days to be kinder to strangers? Is there an easier way to boost happiness in the workforce? Harvard social psychologist Dan Gilbert reckons that there is. Gilbert suggests that all humans have a 'psychological immune system' that helps us feel better and 'synthesise' happiness. No, the answer doesn't lie in sex, drugs and rock and roll. Academics in Europe offer the way forward.

Listening to our favourite music has been a popular way to relax and unwind after a hard day, but researchers at Serbia's University of Nis have found that listening to music releases hormones which cause positive feelings to flood our system and "make us feel content and happy". Beyond lifting our spirits, the research team found that listening to music was effective at improving health. According to Professor Deljanin Ilic the leader of the study: "When we listen to music we like, endorphins are released from the brain and this improves our vascular health."

To test their hypotheses, 74 patients with cardiac disease were split into three groups. Group A was given an exercise routine to follow for three weeks. Group B followed the same exercise routine, but in addition patients were instructed to listen to their favourite music for 30 minutes every day. Group C listened only to music, and did not participate in exercise.

At the end of the programme, participants in Group A were found to have boosted vital measures of heart functionality significantly and increased their capacity for exercise by 29 per cent. The participants in Group B, following the regime of exercise and music, were observed to have increased their exercise capacity by 39 per cent. Curiously, even Group C, who did not undertake any exercise during the trial but simply listened to a daily half-hour of their favourite music, improved their exercise capacity by 19 per cent. Whilst the study was conducted using patients already suffering from heart disease, Professor Ilic's team are confident that the findings are relevant to wider society, based on established knowledge that tells us that regular exercise is good for improving coronary health in healthy adults.

Whilst the conclusions of Ilic's team offers an interesting perspective for the enhancement of wellbeing, their findings are not exactly new. Around 2,800 years ago, simple musical patterns were found to improve the physical performance of Olympic athletes by around 15 per cent.[2] Since then, our fascination with the science of music and its benefit to our wellbeing has continued.

Reactions to music are perhaps considered broadly as subjective but a multitude of clinical studies reveal that cardiorespiratory variables are strongly influenced by music under a range of circumstances. Music has been found to have a positive influence on those suffering from depression (Chan, 2009); shown to reduce hypertension and improve sleep quality (Brandes et al., 2008); reduce heart rate, blood pressure and improve respiratory volumes (Yoshie et al., 2009). In more general studies, slower-paced classical music was found to reduce levels of anxiety in patients awaiting surgery, whilst opera was found to provide effective stress reduction during recovery time for those patients' post-heart surgery.

In their meta-analysis in 2009, Bernardi and colleagues found "dynamic interactions between musical, cardiovascular and cerebral rhythms in humans" (and, yes, tests have been done with 'non-humans' too). Music has also been found to have a soothing effect and reduce anxiety in pigs, dogs and several other mammals.

So, if adding music to the range of workplace wellbeing interventions seems like the way forward, what should we be playing? Professor Ilic at the University of Nis suggests that some genres of music are less effective – such as heavy metal, which is found to raise stress levels – while opera, classical and other kinds of 'joyful' sounds were more likely to stimulate endorphins in the brain. Her researchers observed that music without lyrics was the most effective in creating happiness because music with words "can upset the emotions".

Bernardi's study in 2009 identified that classical music had proportionately greater positive impact than any other genre, and identified that specific musical instruments, including the organ, piano, flute, guitar, harp and saxophone, were all effective at getting the endorphins flowing.

So, it's time to pop another dime in the jukebox and bring the benefit of the sound of music to the workplace. The top five tunes that have been scientifically proven to boost happiness and improve our wellbeing:

1 Bach's Brandenburg Concerto Number 3;
2 Verdi – Va Pensiero;
3 Puccini – Turandot;
4 Beethoven's 9th Symphony;
5 Vivaldi's Four Seasons.

So, will it be beta-blockers or Bach? Pills or Puccini? Surveillance or symphonies? Counselling or classics? Perhaps it's time to take a step back and look at alternatives to the more traditional interventions.

Deploying a wellbeing intervention programme

Given the volume of anticipated benefits for individual employees, teams and departments and the organisation as a whole, by now we might have ready ideas on deploying a range of interventions in the workplace. If you've reached this point in the chapter and haven't been put off by the lack of formal evidence for wellbeing interventions, and have decided that it's a good idea to try some in your business anyway, what should you do?

In order for wellbeing interventions to be effective, the organisation must be ready to implement change. Contextual factors, such as organisational stability or turbulence, industrial relations climate and work demands may all, of course, impact upon the effectiveness of interventions so it's worth thinking ahead before acting.

A realistic intervention strategy will provide a comprehensive wellbeing programme that utilises a range of interventions at primary, secondary and tertiary levels, with both subjective and objective outcome measures. Conditions for successful interventions are complex and will often hinge on local or organisational circumstances – so, interventions should be developed through robust assessment of problems and opportunities, based on robust evidence gained through your own analysis.

Although the maxim 'prevention is better than cure' resonates, we cannot address the source if we do not know it, so it is crucial that the aims of interventions are specifically set out prior to implementation, i.e. to 'reduce levels of stress-related absence' before effectiveness can be assessed. Where interventions are implemented wholesale across an organisation, variability of efficacy may be observed between work groups with some employee groups readier for change, more involved in shaping intervention approaches or simply more (or less) eager to participate in activities.

Bear in mind that an individual at work, and their work environment, are not static objects; there is a constantly shifting dynamic to be considered alongside a potentially endless list of intrapersonal and environmental factors such as personality, social class, cognitive ability, intrinsic and extrinsic motivation, gender, energy, organisational culture and many more. Despite these myriad factors there remains widespread corporate insistence that there is a simple relationship between the individual, their emotional state and their work, which can then be generalised in order to produce sound policy approaches for the masses. Given that work will remain only *one* of these many factors this is a naïve assumption and may also hint at why finding strong relationships between work and wellbeing interventions is so difficult.

Objective work conditions –such as working hours, workload, job grade etc. – have all been studied to check their impact on wellbeing (Spector, 1997). Up to 90 per cent of the variance in the way in which people perceive and rate their work environment is unconnected with aspects of the objective environment, but instead is based on social cues, concurrent mood, attitude and personality and other factors which are not related to the construct being measured. This underlines the important role of corporate culture in influencing the wellbeing or subjective happiness of employees. Prevailing cultures, especially at local levels, may also cause an 'optimising bias' to drift into play – we all like to make ourselves look or feel a little better sometimes, and it's easy to check a box on a survey to do this!

Effectiveness of interventions

Despite growing interest, the relative effectiveness of wellbeing interventions remains difficult to assess, due in part to methodological deficiencies and the lack of adequate evaluation techniques.

Whilst focusing on wellbeing interventions appears to be *en vogue*, especially in times where organisations seek to 'do more with less', how do we gauge their effectiveness? Several theories offer an attempt at identifying links between work and wellbeing. The Cause & Effect Theory, where one thing causes another thing to occur is common, especially the Person–Environment Fit model, where higher levels of mismatch between subjective person and their subjective environment have been associated with higher levels of psychological strain. In the Effort–Reward Imbalance model, both the degrees of fit and imbalance may influence wellbeing, and specifically affect stress. Cause & Effect becomes more sophisticated when introducing moderators such as Karasek's Job Demand–Job Control where the impact of demands on wellbeing depend upon the level of control over the job, or by introducing contingencies, i.e. when two or more factors with a certain high–low combination cause certain outcomes. Whilst this model does have some merit, there remains confusion over to what extent there is an interaction between demands and control, thus the question remains whether the combination of both has an effect on wellbeing that is greater than their additive effect. This model seems too simple and shows a lack of consideration of individual differences but in its defence it does attempt to specify conditions where demands may have positive and negative consequences and this is more than other models have attempted. Generally though, as most studies have centred on point-in-time data, it remains difficult to draw direct correlations between cause and effect and thus gauge the efficacy of our wellbeing interventions.

The Affective Events Theory was the first framework to consider causes and consequences of affect at work by attempting to understand the role of mood and emotion. The theory focuses on structure, causes and consequences of affective experiences at work, viewing events as proximal or immediate causes of affect. In this way, people feel the way they do when at work because of their reactions to events at work, their dispositions and a range of other environmental factors. Longitudinal studies have confirmed the benefits of positive emotions on health and wellbeing, and the 'Happy-Productive Worker' hypothesis, where 'happy' workers are assumed to perform better remains popular with academics and Human Resource practitioners alike.

Although many organisations have established wellbeing programmes, there is little by way of solid evidence to confirm their effectiveness. As interventions often take place at the local level, such as within a team or department, it is difficult to make valid quantitative analysis of their effectiveness. Additionally, often when programmes are implemented, in practice they do not reliably control other influencing factors and are not systematically assessed in the workplace. Fundamentally, measures of wellbeing are typically 'outcomes' and do not reflect the 'wellbeing process'. So, to measure effectiveness we must consider the process of the intervention itself – though process evaluation itself also presents a challenge in that there is a lack of sound theoretical basis upon which to evaluate.

Evaluation of both management and employee perceptions on the effectiveness of wellbeing interventions is vital, as these views may not totally align. In their recent study looking at reducing adverse psychosocial work environment factors, Hasson et al. (2012) found that in more than 50 per cent of interventions, the proportion of employees reporting that a specific change was implemented did not correspond to the amount of change reported by management.

A study by financial services firm KPMG and IDH, the sustainable trade initiative indicates that capital investment in improving working conditions through the application of workplace interventions can pay off within four to 20 months. The report, which followed interventions in 70 organisations and analysed 99 academic studies suggests that positive interventions can reduce employee turnover by as much as 40 per cent, and may also impact the bottom line, with margin improvements of up to 0.4 per cent achievable.

The KPMG/IDH report offers a useful perspective on return on investment – however, typically methods used to analyse costs and benefits will vary between employers and industries, making it difficult to benchmark or generate a shared definition of success. The scientific research that does exist tends to be theoretical, looking at *how* and *why* work and wellbeing may be related, rather than empirical, looking at *what* is actually happening. There are limitations in many of the theories and models that are presented with the aim of improving wellbeing, as most are based on cross-sectional – rather than longitudinal – studies that look primarily at the processes in place in an organisation, and, therefore, do not permit the establishment of causal inference. Most data is gathered using self-report tools, which are open to bias and may misrepresent outcomes – for example, where healthy employees take up interventions a comparison of participants and non-participants may suggest interventions are improving wellbeing more than they actually are. Accordingly, reflecting the true impact of wellbeing interventions is likely to be challenging.

Measuring the effectiveness of interventions is a subjective process, reliant on perceptions of the individual, management and others. There is little published research that explains *how* to assess the effectiveness of intervention strategies and approaches. Although self-reports are prone to the limitations explained above, they *can* help us to learn about people's subjective perceptions of their work, and these perceptions are central to our understanding of wellbeing, however they remain insufficient on their own to truly gauge effectiveness of interventions. Accordingly, choosing to adopt wellbeing interventions in the workplace may come down to simply being a caring and compassionate employer keen to be 'doing the right thing' to help people feel more comfortable, more satisfied, happier and, hopefully, healthier at work and in their lives generally.

Megatrends: the future of health and wellbeing

In this chapter, we've looked at the principal causes of ill-health around the globe in order to consider themes to focus wellbeing interventions upon. We've also explored the less quantifiable side of wellbeing – and as perhaps expected, there appears to be merit in working towards increasing the subjective feelings of happiness in the workplace too. It may now be helpful to look forward to the future. In our changing world, the opportunities and challenges for health and wellbeing at work will certainly include those outlined in this chapter, but, as globalisation continues to pick up pace, new themes will emerge.

Four 'mega-trends', revolving around demographics; lifestyles; economics; and the nature of the work we do, will all present the potential to impact workplace efficacy and personal wellbeing. Let's look at each in turn.

Megatrend 1. Workforce demographics

The world in which we live and work is growing. The global population has more than doubled in the last 50 years. With seven billion of us on the planet now, we can expect this to rise even further to around 10 billion by 2050. This population growth, coupled with the might of globalisation, will continue to spur sociological trends such as increased levels of worker migration and transient labour.

As the population increases, so too does the expected lifespan. The age of an average worker in many developed countries will climb from 39 to 43 by the year 2030. Retirement will be pushed further out, with many of us working well into our 70s and even 80s, presenting myriad issues with regard to safety, health and wellbeing at work. With ageing linked to many common diseases, we should expect that the number of workers with ill-health will rise, as well as anticipating a surge in the number of workers with long-term health conditions.

CASE STUDY: GETTING AHEAD OF THE TREND – SUPPORTING AN AGEING WORKFORCE

A major pharmaceutical firm identified that almost 40 per cent of its workforce at its headquarter facility were over 50 years old. Data analysis revealed interesting themes in sickness absence, prompting the organisation to build a wellbeing initiative especially focused on this age group.

Interventions included a diet and weight loss group which donated money to a local charity – chosen by employees – for every kilogram of weight lost by the group's members. Fruit baskets were placed in more remote locations in open-plan offices, encouraging workers to rise from their desks and move around to eat their '5–a-day'. Finally, the company partnered with a local fitness centre and offered a free two-week pass to all workers over the age of 50.

Results from the campaign were astounding. An 8.5 per cent reduction in absence generated annual savings of £1.2 million. Part of the savings were reinvested in developing the wellbeing programme to include an annual medical check-up for all workers aged 50+, and developing the partnership with the fitness centre to offer subsidised memberships for all staff, regardless of their age.

Megatrend 2. Lifestyle and disease

As we progress into the future our lifestyle choices, work activities, as well as natural evolution will impact our morbidity and mortality. Psychological and psychosocial risks will rapidly rise through corporate risk registers. Musculoskeletal disorders caused by repetitive physical movement and by sedentary work roles will also escalate.

Mental health and physical ability will become the most prevalent health issues for developed nations causing the majority of both short and long-term absence from work. In many countries, whilst the direct costs of disease will be borne by state or governmental healthcare schemes, businesses will also feel the impact. The 'hidden costs of health and wellbeing' such as the indirect costs of absence which may be up to 10 per cent of the organisation's annual salary bill.

In the United Kingdom, around 30 per cent of the total cost burden of disease and disability is directly connected to lifestyle behaviours such as smoking, alcohol, diet and obesity.[3]

Whether catalysed by post-recessional organisational restructures, or increases in disposable income employees will continue to regularly consume excesses of alcohol and smoke tobacco. Despite common sense and state-sponsored campaigns, workers at all levels will exercise infrequently in the future, and the blurring boundaries between work and home will force the favour of convenience foods over healthy eating. Right now, there are more than 550 million clinically obese people around the world. And another 1.6 billion classified as overweight. Plot these people on a world map and you'll find most are in the developed nations. In the United States of America, for example, over a third of the entire population is now considered to be obese, and the country is now home to the highest rate of obesity in the developed world.

But the story gets worse. As countries shift from 'developing' to 'developed' status, workers will look to the Western world and emulate lifestyle behaviours in attempts to fit in, fast-tracking these negative lifestyle trends and exacerbating the problem across the globe.

For developed nations, coronary heart disease (CHD) will continue to be the biggest health cost to employers. In 2012, the disease accounted for some £4 billion in costs per year in the UK. Current projections show that CHD will continue to increase until 2030. By that time, the costs of the disease could be treble those of today. The 'hidden costs' typically considered in work-related accident cases – such as productivity losses, absence, induction, training, supervision and salary costs of replacement labour, additional line management administration – will become more apparent as organisations become aware of the true cost of poor health. With more heart attacks occurring on a Monday than any other day of the week, employers who have not yet noticed the impact of the disease, in the future certainly will.

Beyond the known diseases, the World Health Organization suggests that the greatest future health risk to life satisfaction and productivity is *dysthymia* – a sharp and distinct loss of energy that hampers the maintenance of valuable social relationships, causes individuals to doubt their own certainty and dents personal feelings of happiness and contentment.

Whether caused by dysthymia, longer working hours, stress, physical tiredness or lack of interest our lifestyles are killing us quicker than ever before. In 2012, there were 2 million deaths every day due to sedentary lifestyle. Two million. Every single day. And this figure is only expected to increase between now and 2050.

So, what's the cure? An easy route to improving health is, as we all are aware, exercise. Although it fast becomes a circular construct, on the face of it, the idea that healthy workers are happier and more productive makes good sense. Gently encouraging employees to regularly engage in moderate exercise benefits everyone.

If your workplace doesn't lend itself to the creation of running or walking trails there are many other options. Running clubs, sponsoring employee places in the local 5k footrace, subsidised subscriptions for the gym, family fun days . . . the list of possible interventions that are straightforward, easy to implement and bring almost instantaneous results goes on and on.

Financial incentives may be effective in changing certain behaviours more than others. Research from the United Kingdom (Jochelson, 2007) suggests that the most successful incentives are those that focus on simple behaviours such as keeping appointments, or performing a prescribed activity at a certain time. Whilst incentives may help people to achieve personal behavioural change goals, there is clearly a risk of the new positive behaviours lapsing when the incentive is removed. Incentive schemes may not be sufficient on their own to counter the influence of social pressures (for example, to drink alcohol or smoke), personal habits or psychological dependency. The *British Medical Journal* reports that the involvement of a qualified clinician or counsellor can often significantly increase the chances of sustaining longer term behavioural change.[4]

Financial incentives are not widely utilised in workplace wellbeing programmes, however in the United States, where organisations are duty-bound to provide health insurance for full-time employees, a combined 'carrot and stick' approach – using financial rewards and penalties – is often used to encourage employees to choose healthier lifestyles.

CASE STUDY: GETTING AHEAD OF THE TREND – WALK THIS WAY

A multinational financial services company recognised and appreciated the benefits of regular exercise. As they constructed their new global headquarters on a brownfield site just outside the city centre, they wove walking trails through the grounds of their campus. Every morning, lunchtime and throughout the day, executives, managers, clerks and assistants could be found strolling in the fresh air. There was plenty of room for those wishing to pick up speed, and a works running club was soon formed by the employees themselves as the loops provided an adequate choice of distances to train upon.

As the company settled into their new location, the trails evolved from being part of the site, to becoming part of proactive wellbeing campaigns. The organisation began to actively encourage people to participate in 'walking meetings', and in fine weather small groups of employees could be observed wandering, stopping to cluster and discuss more delicate points.

The success of the trail paths, installed at little cost and requiring rare maintenance, enabled the company to free up space previously allocated as meeting rooms, essentially gaining more room for offices. Employees meanwhile, through their feedback in engagement surveys, attested to the importance of the paths, advising that they felt fresher and more alert, higher levels of motivation and generally more effective and less stressed during the working day.

Megatrend 3. Economic factors

The old adage 'money makes the world go round', whilst still relevant for many, is being turned on its head as more and more organisations 'go round the world to make money'.

As the world battles to emerge from the financial apocalypse that began in 2008, tentative recovery is the order of the day. Measured strategy for global expansion, including the capitalization on markets with reduced labour costs through corporate relocation may provide economic incentive for growth in uncertain times.

Globalisation, the process of integrating business across international arenas, brings benefits in terms of productivity, efficiency, cost and proximity to the customer. But it also presents challenges from an employment perspective. Looking back over the last century we can see that the strength and direction of the economy – whether in our base location, or those countries where expansion is intended – and the health of the workforce are strongly linked. We know that healthy workers are more productive. On the other side of the coin, changes in the local economy can deliver intense impact on the health and wellbeing of workers.

The financial downturn, coupled with an underlying insistent desire for globalisation, catalyses strong feelings of insecurity around employment – for those who have found themselves on the receiving end of redundancy, and for those that have survived the cutbacks. Such fears lead to a marked rise in unhealthy lifestyle choices and behaviours as increased usage of alcohol, smoking or recreational drugs are employed as coping mechanisms.

CASE STUDY: GETTING AHEAD OF THE TREND – PAY TO WIN

A manufacturing company based in Europe devised a scheme to encourage pregnant women to give up smoking by offering them a 50 Euro supermarket voucher if they passed a weekly carbon monoxide breath test. More than 80 per cent of those registered in the scheme stopped smoking. Anecdotal evidence revealed that several of the women's partners also gave up smoking at the same time, despite not being part of the incentivised scheme.

A major American food group operated a similar scheme with the objective of convincing staff to reduce their levels of alcohol intake. In a six-month pilot programme 90 employees across five locations committed to being 'alcohol free' by consuming no alcohol during the week, or on Sunday evenings. Employees were required to blow into a breathalyser each morning and have their results recorded. Those who attained an 'alcohol free' status at the end of the programme could choose from a selection of rewards which included holiday and gym vouchers, cases of Champagne, new bicycles, television sets and family activities. Interestingly, none of the employees chose the Champagne for their prize.

Recessions, double-dips and downturns naturally cause organisations to review their spending, yet, rather ironically in such harrowing times, investment in workplace health and wellbeing programmes is often identified as an 'easy cut' by less forward-thinking CFOs. Blind eyes are turned to the 'alternative methods' utilised by employees to 'see them through' the corporate rollercoaster rides.

Globalisation will also serve to increase the demand for the most highly skilled workers. On the face of it a positive note, but disparity between the skills required and actual skill level available will lead to the employment of under-skilled or inexperienced individuals being placed into roles which are simply beyond them. Silent suffering will become the norm for many of these individuals, their feelings of stress, exhaustion or inability muzzled as the thrust and excitement of global integration and expansion acts as a metaphorical gag, preventing them from speaking out and sharing their concerns.

As the road to globalisation unfolds and international boundaries are broken down, opportunities for working in new geographies will become standard, swelling the ranks of those who relish the 'expatriate' lifestyle, or who see the value in broadening their experiences. Increased worker transience, from relocated fixed-term-contract working, to the advent of the 'mobile manager' constantly travelling the globe building the business or troubleshooting tribulations, will add weight to the balance of work and wellbeing. As more and more cities around the world become interlinked by low-cost air travel and improved local infrastructures, working away from home will become the norm for many. But the price of constant corporate entertaining and hotel lifestyles will quickly take its toll on the 'have-suitcase-will-go' cohort.

In the future, the pressure for 'good jobs' will intensify, resulting in it becoming commonplace for employees to attend work even when they feel unwell. Whether fuelled by job insecurities, career development or 'saving face', the increasing trend of 'presenteeism' will overtake traditional sickness absence, and lead to costs for employers of more than twice that of regular absence.

Megatrend 4. The nature of work

The three preceding issues serve to collectively drive a fourth mega-trend that will affect worker wellbeing in the future. Changes in demographics, economies, cultures and technologies will collide and fundamentally adjust the way we work across the world. In many developed countries our reliance on information technology, cloud computing and remote networks will continue to grow, synchronously with the continued escalation of the knowledge economy, prompting an evolution from hot-desking to 'out-locatiing' of workers as organisations reduce their required office space and save costs. The lines between home and work will blur beyond recognition as globalised organisations become 'open all hours'. The old 9–to–5 pattern will vanish as the working day extends around the clock, punctuated with self-directed micro-pauses for laundry, childcare, leisure and social activities.

CASE STUDY: GETTING AHEAD OF THE TREND – WELLBEING ON THE MOVE

A global manufacturing company, in the midst of an intense programme of strategic acquisitions, had a group of managers on an almost permanent cycle of moving around the world conducting due diligence assessments, engaging with local workforces and stakeholder groups and building and implementing integration plans. All of which needed to be coordinated via conference calls with headquarter facilities in the United States and Europe.

When several of these managers began to report in sick with exhaustion, or prone to regular colds and fevers brought on by the frequent travel and long hours, the company realized something needed to change.

Not sure on what exactly to change, the company organised that prior to departing on these trips, the managers would meet with a travel doctor who provided not only the necessary vaccinations but also advice on maintaining wellbeing during the periods away. The doctor counselled the staff on the challenges of working across time zones, offering tips on combatting jetlag, healthy eating ideas, providing simple bodyweight exercise routines that could be done in their hotel rooms and installing guided meditation applications on their company smartphones.

The absences previously experienced vanished as the management team became more aware of how to help themselves stay fit and well during trips. Their positive feedback encouraged the company to create a Travel Wellbeing Pack which is now given to all staff before they begin a period of working away.

As heavier industries and manufacturing operations are outsourced to developing nations as part of corporate globalisation campaigns, these 'new' activities will bring new health and wellbeing risks to workers. From hazardous substances to nanotechnology, high-speed process automation to manual handling, repetitive movement and unfamiliar work environments, each will present unique challenges as accelerated learning is encouraged in order to satisfy growth and customer demand.

Whether developed or developing countries, wrapped around these shifts in the nature of work, the amplification of Corporate Social Responsibility will continue to shine the spotlight on the employers' obligation to offer 'good quality work' – employment which optimises the positive impact of work on health and facilitates higher levels of wellbeing among employees and results in lower incidences of physical or mental illness.[5]

CASE STUDY: GETTING AHEAD OF THE TREND – THE GOOD LIFE

A global FMCG company implemented a 'Healthy Working Life' initiative aimed at maximising performance and increasing employee motivation and job satisfaction. The programme included self-service intranet-based resources, online learning modules and one-to-one coaching via telephone calls with trained advisors.

In addition to boosting employee morale, engagement and satisfaction, the programme also positively impacted the bottom line, reducing absence rates by around 2 per cent. Return on investment was calculated as a ratio of 4 : 1, and generated savings of over five million USD.

THE NAKED TRUTH: HEALTH, WORK AND WELLBEING

The mystery that has historically shrouded definition of the term 'wellbeing', has begun to lift – and not before time. However, challenges and doubts persist around the effectiveness of wellbeing interventions. The few scientific theories that exist remain lacking rigorous development, empirical analysis or systematic testing.

In times of scarce resources, 'off the shelf' packaged interventions that are offered widespread to the entire workforce may be presumed to be most effective investment, given that they reach a wider audience. A 'one-size-fits-all' approach rarely benefits everyone.

Health and wellbeing behaviours are complex, determined and influenced by more than just an individual's level of knowledge. Wellbeing programmes need to reflect this complexity, so to be more effective in influencing change, a range of strategies for informing, educating and involving employees is necessary.

The good news is that as the boundaries between work and domestic life have become blurred, it's become more socially acceptable for employers to take an interest in the health and wellbeing of their workforce and offer encouragement to assist employees in leading healthier lives.

The workplace provides a useful environment to raise awareness about disease, health and wellbeing; to promote exercise, nutrition and healthy living; and to provide opportunities for confidential discussion. These may all enhance the work experience.

To be effective, health and wellbeing interventions should reflect the changing nature of work, the way it is organised and the role it has in modern life. Robust risk assessment and two-way dialogue will help to identify key issues.

It's a basic truism that most people are motivated by what makes them happy. Reviewing the level of workplace contentedness is a useful step to identifying new perspectives on enhancing wellbeing that can pay dividends both at work and at home.

Organisational culture plays an important role. A positive culture that actively engages employees through open dialogue is an excellent foundation for the development of workplace wellbeing.

GET NAKED: HEALTH, WORK AND WELLBEING

1. What are the main causes of absence in your workplace? What trends or patterns do you see when you review absence data?
2. How do you measure employee engagement in your workplace? Do you consider employee wellbeing, happiness and job satisfaction in these measures?
3. If you currently implement wellbeing interventions in your organisation what is the desired aim of these programmes? How do you measure their effectiveness? Are you measuring 'outcomes' or 'process'?
4. How does your organisational culture encourage positive emotions or 'happiness' at work? Where can you create opportunities to boost workplace happiness in your organisation?
5. Of the four 'future mega-trends' in health and wellbeing, which do you think is most important to your business? What plans do you have in place to manage these challenges?

Notes

1. See research by Jessica Pryce-Jones, iOpener, Institute of People and Performance. Also Baron, 1990; Horn and Arbuckle, 1988.
2. Hans-Joachim Trappe, Department of Cardiology and Angiology, University of Bochum, 'Music and clinical health' paper to the ESC Congress, 27–31 August 2011.
3. 'European Health for All' database, Copenhagen, WHO Regional Office for Europe, 2008.
4. Lancaster, T., Stead, L., Silagy, C. and Sowden, A. (2000) Effectiveness of Interventions to Help People Stop Smoking. *British Medical Journal*, 321(7257), 355–358.
5. Vaughan-Jones, H. and Barham, L. (2009) *Healthy Work*. London: BUPA and Boyce, T., Robertson, R. and Dixon, A. (2008) *Commission and Behaviour Change: Kicking Bad Habits Final Report*. London: The Kings Fund.

17 Sustainability

Sustainability is a corporate management concept that offers an alternative to the traditional growth and profit-focused business models.

Whilst recognising that growth and profitability are vital, sustainability directs the organisation to follow societal goals at an operational level, specifically those relating environmental protection, social justice and equity and economic development.

In this chapter, we'll explore why sustainability is important, how initiatives need to be managed, why avoiding 'green-washing' is crucial, how best to collaborate, and why engaging on sustainability is vital to organisational success.

According to *A New Era of Sustainability* (2010), a global study by the United Nations, 93 per cent of CEOs now believe that sustainability is an important factor to an organisation's future success – yet, only 17 per cent feel confident that they are ready to face its challenges.

With the natural resources of the world continually plundered – and now beginning to dwindle – it means that values and behaviours matter more now than ever before.

Organisations are now facing significant scrutiny about their operational activities – with a hungry world press constantly seeking out stories to break about corporate foul-play, unethical behaviour, environmental pollution, workplace accidents and dangerous lapses within their supply chains.

News stories, accident investigations and external governance regularly discover considerable anomalies around what organisations claim they stand for, and the reality of the situation.

When the automotive company Volkswagen was discovered to have cheated the emission-testing process for its diesel cars in 2015, the fallout was enormous. Having misled the US public as to the environmental impact of its products, it transpired that the owners of some 11 million Volkswagen cars were, in fact, generating the same amount of pollution as the greenhouse gas emissions of every British vehicle, factory and power station combined. The resultant scandal wiped billions off Volkswagen's share price, losing the business approximately a quarter of its market value in a less than a week.

Choose any sector, and you'll quickly discover that sustainability – including the disciplines of safety, health and environment – has become an increasingly important part of doing business.

For organisations to balance their financial, social and environmental risks, obligations and opportunities, sustainability needs to move from being an add-on to truly being part of culture, being essentially 'the way we do things around here'.

Organisational culture is now being recognised as a fundamental part in the shift toward sustainability. However, despite myriad corporate sustainability reports that describe sustainability as "the way we do business," most business leaders lack a clear understanding of *how*, *where* and *why* they should embed sustainability in their day-to-day decisions and processes.

After all, how does an organisation go about maximising efforts and deepening our impact, so that our good intentions turn into actual, positive change?

What gets measured gets managed

Before embarking on a convoluted sustainability initiative, it is usually a good idea for organisations to start simply by evaluating their own impact – and getting their house in order.

It is useful to begin with a formal audit or assessment, comprising of everything from recording the amount of recycling taking place within the organisation, to examining utility bills, to looking how much food is wasted by the staff canteen.

Once a baseline has been established, it will then be easier to track where opportunities for improvement lie within the organisation. Typically, the most logical – and often, easiest – first step towards sustainability an organisation can make is to locate systemic bad practice and address this. From an organisational perspective, then, sustainability is about investigating the effects of everyday operations and ensuring that the good consequences outweigh the bad ones.

Starting simply is also useful in raising awareness of the subject in a non-threatening way, before engaging with workers, and getting sustainability talked about across departments. It's also important for leaders to put in place a system that makes it easy for workers to make choices that enhance sustainability.

Frontline staff are in a unique position to give valuable insight into how to reduce energy use and waste. As we've discussed in other chapters of this book, organisations risk losing significant value when they underestimate the potential input of workers to broader organisational activities, as some of the best and most innovative ideas comes from the individuals actually engaged in everyday tasks.

Stakeholders

Stakeholder engagement is a key part of any sustainability strategy – yet, it is often something that is under-prioritised and badly managed. Engagement of stakeholders is at the core of any effective strategy, as an understanding of stakeholder perspective will help forge a united front and will – hopefully – mean that everyone is pulling in the same direction in the future. Also, if things prove problematic, it is better to find out at an early stage.

Sustainability is not about philanthropy or charity. Nor is it something that just about generating good Public Relations. (Though, there does seem to be some general confusion about this within a number of industries.) In reality, sustainability is about keeping the organisation afloat. As such, it is important that sustainability is taken seriously – and has the backing of the Board of Directors.

The Directors have oversight of the functioning of every aspect of the organisation – but, also, crucially, represent the interests of the major stakeholders.

To engage effectively with stakeholders about sustainability, Directors need to embed it into board practices – and tailor messages to stakeholders about it that will make them sit up and take notice.

For example, in many organisations the license to operate will be a subject of great concern at the Director level. Boards can thus stress the need for sustainability by pointing out how any behaviours by the organisation deemed risky and immoral, could be potentially disastrous to license renewal.

Internal education with regards to sustainability initiatives is often more complicated external engagement. The concept that 'the only business of business is business' is a persistent one in every industry – though, as we'll see, no longer a very accurate one.

To build confidence amongst stakeholders, it's a good idea to identify – and industriously advertise – any 'quick wins' throughout the organisation, as this galvanises support by clearly demonstrating that progress is real and happening.

External engagement – airing your clean whites or exposing your dirty greens?

Though sustainability is generally perceived as a positive attribute – and, accordingly, the prevalence of sustainability drives has increased rapidly over the last few decades to meet consumer demand – due to media coverage of 'greenwashing' a great deal of scepticism has built up around it.

Greenwashing describes organisations using branding, PR or marketing collateral to deceptively promote the idea that an organisation's products, aims or policies are environmentally friendly.

It can take many forms – from rebadging products to promote the idea that they are more eco-friendly (think back to when all those green washing-up liquid bottles suddenly started appearing on supermarket shelves) to the much more serious use of advertising to mask activities that are extremely dangerous to people or the environment.

A good example of the more serious type of greenwashing took place in Malaysia in 1993. The British chemical company ICI published a full-colour newspaper advertisement for one of their new products under the headline: "Paraquat and Nature Working in Perfect Harmony". The copy within the advert described the new pesticide Paraquat as "environmentally friendly" – as well making other various spurious claims about its general wholesomeness – even though the company must have been aware that it is a highly-toxic herbicide that had already poisoned many thousands of workers in Malaysia alone.[1]

In order to combat perceptions – or, indeed, accusations – of greenwashing, it's essential that organisations define their sustainability programmes, and that that actions and outputs within these programmes are measured, accurate and stand up to scrutiny.

Many organisations leave the sustainability strategy in the hands of the Public Relations or marketing teams – and, consequently, their press releases, marketing documents and webpages suddenly become filled with boastful statements designed to make the organisation sound excellent – rather than an accurate representation of the facts. This is not a complaint about PR or marketing – after all, this is largely what they are hired to do. However, when it could hugely damage an organisation's reputation and perceived trustworthiness, it is dangerous.

One common way of greenwashing is often referred to as 'the sin of omission' – it is to highlight one positive feature of a product or service – and quietly disregard its other qualities.

Fast-food giant McDonalds were recently accused of this, when they – literally – tried to paint themselves 'green'. In all mainland European territories, the iconic yellow-on-red McDonalds logos were switched for yellow-on-green ones. The organisation took this slightly extraordinary step, following a high-profile switch to use biofuel made from leftover grease in its fleet of lorries, and a move to use recycled paper in its takeaway bags. However, many campaigners – and, indeed, customers – seemed unsure if an organisation whose entire business model is based around disposable packaging could really claim to be quite so environmentally friendly.

Ensuring all sustainability statements are accurate and representative is essential. Claiming, for example, that a product 'is made of 100 per cent recycled materials' – when, in reality, it's only made from 97 per cent recycled materials might feel fine in promotional literature – as though the organisation is simply 'rounding up'. However, if it's discovered that your claims are embellished – and, therefore, not actually true, it might be very damaging. Exaggerating any sustainability claim – even by a negligible amount – could be detrimental to an organisation's reputation and lead consumers to believe it dishonest.

When promoting their sustainability strategies, organisations need to strive to be as transparent as possible. As such, it makes a good deal sense to try to 'pre-empt the argument' by backing up claims, publicising data and being open to industry (and public) scrutiny.

It is also beneficial for organisations to have their products and services independently inspected by outside auditors, and to obtain any relevant certifications – as this will naturally add authority.

Collaboration

Organisations need to collaborate, because businesses and markets no longer operate in isolation. Multiple stakeholders – shareholders, suppliers, regulatory bodies, local communities – now have a greater impact than ever before on how organisations work.

Simplicity is key to successful collaboration. The most effective collaborations focus on single issues rather than a broad range of aspects. Take, for example, the recent collaboration between Marks and Spencer and Oxfam, designed to encourage the British public to do more clothes recycling. The high-street clothing store and the charity teamed up with a simple proposition that meant that whenever an item of M&S-branded clothing was donated to an Oxfam shop, the individual donating it received a gift voucher giving them money off from their next shop in Marks and Spencer. It was a huge success and has opened the door to many other retailer/charity collaborations.

Collaboration is widely regarded as essential if organisations are to genuinely attempt to address the global challenges at the rate needed – but it is not always easy to implement. Forging and maintaining relationships that are essentially 'extra-curricular' can be difficult for any business – but can be especially problematic if that organisation is a competitor. Though they might seem to be the obvious choice to partner with an organisation within your own sector, it can often be tricky to truly collaborate with a business with whom you are also routinely battling for market share.

When looking for the opportunities for collaboration and potential partners, organisations need to think broadly. It is a good idea for organisations to look beyond their own industry or sector – but also *within* their own organisation.

A sense of purpose

In 2011, a Gallup poll of 2,341 employed adults in the United States showed the surprising fact that there is almost no correlation between income level and worker 'job satisfaction'.

Though remuneration is a motivator during the job-seeking process and when it comes to worker advancement, it seems to be ineffective at maintaining worker engagement.

In his book *Drive*, Daniel Pink draws upon a vast quantity of behavioural research to determine what motivates workers.

Pink asserts that there are three basic forms of motivation:

1 Extrinsic or External Motivation – motivation based on rewards and punishment;
2 Internal or 'Creative Motivation' – self-motivation driven by internal desire;
3 Purpose-driven motivation – which Pink describes as the most powerful motivation.

Purpose-driven work is now increasingly linked with job satisfaction – which is why organisations that implement sustainability initiatives often report significant increased performance.

In the article *Does Sustainability Change the Talent Equation?* (Brokaw, 2009), MIT Sloan Management Review claims that there is a growing trend in new graduates to actively seek out job roles in progressive organisations where strong sustainability drives. It states: "People are hungry for the opportunity to work professionally in a way that is consistent with building a sustainable world instead of one that undermines it."

Consequently, there is a growing importance placed on jobs that have a sense of purpose – and which tap into the passions of workers. If organisations want to attract – *and retain* – talent, offering potential employees good salaries is no longer enough.

Sustainability and competitive advantage

It is not uncommon for sustainability drives to be dismissed out of hand by business leaders as little more than decorative endeavours – a bit of window dressing which, whilst making good copy in press releases and marketing campaigns, is essentially a distraction from the proper business of the organisation.

Indeed, leaders might be forgiven for thinking so – after all, it seems logical that anything that uses up valuable time and resources without actually contributing to the core business is unlikely to add value.

However, this doesn't seem to bc borne out by the facts. According to research by Goldman Sachs, share prices of organisations that are leaders in Environmental, Social and Governance (ESG) policies outperform others by around 25 per cent each year.

Furthermore, Oxford University's Smith School of Enterprise carried out a study of more than 190 academic studies on the impact of ESG policies on performance in 2014. The report claimed that 80 per cent of the data analysed showed that there was a proven 'positive correlation' between sustainability and stock market performance. By contrast, organisations that routinely violate environmental regulations experience a significant drop in share price.

Becoming sustainable is fast developing into a strategic decision for every organisation. In the most recent annual *Sustainability & Innovation Global Executive Study* (Kiron, 2015), a study sponsored by the corporate intelligence organisation SAP and energy giant Shell, 70 per cent of organisations claimed that, rather than negatively impacting their competitiveness, sustainably had actually helped increase performance.

The research also showed that of the 2,800 respondents interviewed – made up of international business leaders – approximately 31 per cent stated that sustainability was actually growing their profits, with about 70 per cent claiming to have put sustainability permanently on the management agenda.

In many parts of the world, it's becoming increasingly likely that regulation will drive organisations to into becoming more sustainable. As a result, those business leaders that were early-adopters and quick to see the benefits of becoming more sustainable, will have put themselves at an immediate competitive advantage.

Creating shared value

In 2011, *Harvard Business Review* published an article by Michael E. Porter entitled 'Creating Shared Value: Redefining Capitalism and the Role of the Corporation in Society'. Porter, a leading authority on competitive strategy at the Harvard Business School, provided research that showed that there were clear links between business strategies and Corporate Social Responsibility (CSR).

The following year, Porter co-founded the 'Shared Value Initiative,' the central tenet of which is that organisational competitiveness and the health of the community in which it is located enjoy a symbiotic, mutually dependent relationship. Essentially, for an organisation to have sustained success, it must create value not just for its shareholders – but also for society.

One of the first multinational organisations to adopt a 'Creating Shared Value' (CSV) approach, was the Swiss food and drink giant, Nestlé. The organisation decided that the best way to proceed would be by focusing on the social issues that they themselves were in a strong position to influence. As such, the organisation concentrated on three areas – nutrition, water and rural development – which it saw as core to their 'business activities and vital for their value chain'.

In 2016, Nestlé received vindication for their CSV activities when they received industry-best scores for all three areas of sustainability – economic, environmental and social – by the Dow Jones Sustainability Index (DJSI), a globally renowned independent standard that measures the performance of the world's largest 2,500 companies.

Though the aim of Nestlé's CSV activities was to ensure a sustainable and growing business, the organisation – which has been dogged by controversy over the course of its long history – has been bombarded with a wave of positive publicity, resulting in a much-improved global reputation.

Sustainability as a core value

Organisations today operate in a world of finite resources and growing competition.

On the face of it, this would suggest we are living in an increasingly dangerous climate, likely to result in big business grabbing whatever it can – in an effort to eliminate their competitors. However, in the most part, there is something that is standing in the way of multinationals carving up the planet's dwindling natural resources and then selling them on – and that is the consumer.

Consumers like sustainability. Of course, they do – why else would businesses go out of their way to promote it? For what other reason would the concept of 'greenwashing' exist?

Back in the heady days of the late 1980s, when the ice cream retailers Ben & Jerry's published the first-ever 'social audit' report, it was met with some palpable confusion – and promptly dismissed by most business leaders. Less than 30 years down the line, if an organisation isn't producing annual reports on social and environmental accountability, its perceived as backward and neglecting its social duties.

Big business will always bow to social pressures in the interests of sustainability (in many cases, it's their own sustainability they're thinking about) – and the consumer wields an enormous influence.

This pressure by the consumer has led to organisations throwing vast resources into building more efficient processes, creating more eco-friendly products and rolling-out vast waste-reduction campaigns. It has even led to massive multinationals like the US conglomerate Walmart sending shockwaves through its own supply chain by demanding that all of their suppliers evaluate the full 'environmental cost' of their products – at the risk of being dropped by the retailer.

These same pressures have also had a knock-on effect on how many organisations operate. For example, Organisational Safety and Health (OSH) has stopped being simply a regulatory hurdle that needs to be met. Many organisations pushed OSH further – discovered how it positively impacted their bottom line – and, in many cases, 'safety' quickly became a visibly-promoted core value.

With fewer instances of absenteeism and less money spent on staff turnover and settlement pay-outs, this shift in focus has ultimately benefitted not only individual workers and organisations – but also society.

The obvious correlation between levels of worker health, safety and happiness and improved performance, has become largely recognised – leading to many organisations now pushing worker 'wellbeing programmes' to the top of their agendas.

However, though it is now common for big business to support sustainability measures, when the great majority are doing so simply to fall-in with government regulations, or simply submitting to societal pressures, it still often feels like what they are doing little more than paying lip service or creating PR opportunities.

For example, the disappointing way in which the multinational organisations associated with Rana Plaza acted in the wake of the factory collapse might be seen as more telling than any of the words on their promotional literature or corporate websites.

Fortunately, with more and more organisations beginning to see that – like building health, safety and wellbeing into their operations – implementing sustainability initiatives is actually good for business, it is likely to endure.

However, like safety, sustainability needs to be a core value – before it's too late.

THE NAKED TRUTH: SUSTAINABILITY

Over the last few decades, broad societal concerns about sustainability have grown steadily to become a major theme in the corporate world.

It is perhaps not surprising, that protecting the natural resources that our economy and society depend on, has finally found its way on most progressive business agendas.

Though they are now more prevalent, sustainability initiatives are not always easy to implement. Strong governance is required to drive them – and, to succeed, it is essential that the key concepts are embedded into every organisation from the top – and persistently and visibly promoted.

Environmental, safety and social performance indicators also need to be continually assessed – and with the same level of scrutiny and vigour as those for financial performance.

Crucially, a successful sustainability campaign is likely to depend on the amount of attention paid to relationships with stakeholders – including frontline workers – through ongoing engagement.

However, though often problematic – sustainability been seen as essential. In the last 20 years that the Harvard Business School has been monitoring corporate responsibility, organisations that have embraced sustainability – known by them as 'high-sustainability companies' – have consistently outperformed those that have not.

Though many business leaders still cling to the misguided belief that sustainability erodes organisational competitiveness, squanders resources and destroys shareholder value, research has proved this to be inaccurate.

Modern society has expectations about sustainability – and if these expectations are not met, it quickly becomes a source of significant competitive disadvantage.

As societal pressures intensify in the coming years, the performance advantage of high-sustainability companies will naturally rise too.

GET NAKED: SUSTAINABILITY

1 How important is sustainability to your organisation? Look around your workplace: what measures have been taken to reduce energy use and waste? What could you do to further improve things?
2 How often does sustainability come up in day-to-day conversation? Is it routinely mentioned in meetings? Do you and your co-workers mention it on a regular basis? What does this tell you about your organisational culture with regard to sustainability?
3 Do you think that sustainability is recognised as a core value? Why do you think this? When and how do leaders actively promote the concept and demonstrate that it is a value to your organisation?
4 Create a stakeholder map for your organisation. Where are the opportunities for you to share some sustainability 'quick wins' with your stakeholders?
5 Who could you collaborate with in order to promote the sustainability agenda? Think both within your organisation and externally – within and beyond your industrial sector.
6 There's a drive towards people wanting to do work that has purpose and meaning. How could you share what your organisation does in a way that allows your staff to understand their role in ensuring sustainability?
7 How does your organisation create shared value and contribute to the societies and communities in which it operates? What further opportunities are there for you to consider?

Note

1 Paraquat, which has an extremely high oral toxicity, has since been described by the World Health Organization as one of the world's 'most deadly chemicals'.

18 Corporate social responsibility

Recent academic research and marketplace studies reveal that 'doing the right thing' generates positive impact towards the organisation – in terms of reputation-building, employee morale and stakeholder engagement.

In this chapter, we'll explore the concept of Corporate Social Responsibility and its evolution in recent years. We'll consider how an organisation can build a strategic approach to getting CSR right by reviewing two simple models of CSR and look at examples from leading organisations in order to provide some inspiration for you and your business.

In April 2013, the Rana Plaza, a building containing several clothing factories in Bangladesh, collapsed. After 20 days of searching by local authorities, the death toll was finally recorded at 1,127. It was South Asia's worst corporate accident since the Bhopal disaster in 1984.

The building – which, it later transpired, had been built on top of a pond that had been hastily filled with sand – collapsed entirely.

Further investigation turned up the fact that only five of the factory's eight storeys had actually been granted planning permission.

Finally, it was discovered that the day before the disaster, local police had ordered the building evacuated, after staff had reported large cracks spreading up through the walls in the lower levels of the building. The factory's owner, Sohel Rana, disagreed with the police assessment of his building – and quickly ordered his staff to return to work, on penalty of forfeiting a month's salary if they dissented.

Rana was later arrested trying to flee the city, and, as a result of the worldwide coverage afforded the disaster, Bangladeshi Prime Minister, Sheikh Hasina, was later forced to concede that 90 per cent of the country's commercial properties conformed to no building code.

The International Labour Organization (ILO) stated: "Horror and regret must translate into urgent and firm action. Action now can prevent further tragedy. Inaction would mean that the next tragedy is only a matter of time" (Reddy, 2013).

Director of Policy and Research at the International Institute of Risk of Safety Management, Barry Holt, at the time said of the Rana Plaza disaster: "Inevitably, fingers are pointed at the building developers for exploited loopholes in planning laws, and at legislators for not introducing and enforcing stricter planning controls."

However, are they the sole cause of the disaster? Whilst it's true that the Bangladeshi government must be held accountable for their systematic failure to enforce their own national building codes – what responsibility should be meted out to the multinational organisations whose clothing orders were actually being processed in the factory?

Over the last decade, Bangladesh has seen a massive growth in its garment industry – to the point that it is now the world's second-largest exporter of clothing after China.

Many Western fashion brands have flocked to the country, finding its combination of low-cost and accommodating workers desperate for paid work, quick turn-around times and very little government intrusion an attractive prospect. However, in the wake of Rana Plaza, these same organisations suddenly found themselves condemned internationally for exploiting poverty-struck workers and exhibiting a callous disregard for their safety in the name of profit.

Two organisations that attempted to make reparations following the building's collapse were the Canadian clothing company Loblaw, owners of the fashion brand Joe Fresh, and the low-cost high-street label Primark, both of which quickly promised compensation to the victims' families.

Moral management

It is increasingly commonplace for multinational companies to pay out money in the name of CSR – and has been since the 1990s, when international media exposed the sporting goods giant Nike's use of sweatshops as suppliers – and highlighted a number of associated child labour issues.

Nike had been using sweatshops in South Korea and Taiwan to produce sporting goods since the early 1970s. However, when the local economies improved, they decided to look for cheaper labour in China and Vietnam – which also prohibited labour unions.

Many brands have moved their production facilities to developing countries – fortunately, few have exhibited the same degree of single-minded focus on profits as Nike. The problem with stretched supply lines and outsourcing to workers in factories in far-flung parts of the world, is that the sites are rarely visited by members of the parent company – and so, usually, very little is known of the working conditions other than what might be told to them by their owners.

Though the term CSR has traditionally been used fairly indiscriminately to discuss moral corporate issues relating to economic and legal accountability, in many instances it comes down to certain hard practicalities. However, in this case, it would seem as though none of the multinational fashion companies working out of Rana Plaza had even gone as far as asking to see a valid building certificate – which might have been considered a fairly basic check.

Recently, the definition of CSR has been opened up to include more ethical and humanitarian responsibilities. Now it is usually seen as an organisation's obligation to its stakeholders – which might include customers, workers, suppliers, investors and the communities that surround it.

Stakeholders have varying needs to be met. Whereas a customer's greatest concern may be the safety of a company's products, an employee's need might be for a fair wage and safe working conditions. An investor may be concerned with profits and the bottom line, while the community may care about a business limiting the pollution it causes. Thus, corporate social responsibility means maximising the good and minimising the bad effects your company has on these stakeholders' diverse interests.

In *The Pyramid of Corporate Social Responsibility: Toward Moral Management of Organizational Stakeholders*, business management author and professor Archie Carroll claims that CSR is made up of four interconnected responsibilities – economic, legal, ethical and philanthropic – and he believes that true corporate social responsibility cannot be achieved without meeting all of these elements. Carroll's pyramid is widely used to explain the main areas that a business's duties to its stakeholders fall under.

Whilst in the aftermath of Rana Plaza and other such disasters the principles of equating big business with CSR seems very attractive to most consumers, when it is not directly associated with a particular organisation or event, consumers seem to lose sight of its usefulness – and the results become confused.

Research by the social enterprise *Give as you Live* showed that the vast majority of consumers thought that organisations instigating humanitarian and charity campaigns were a good thing – and that these kinds of activity would be liable to foster loyalty and support. In fact, a massive 79 per cent of British consumers polled believed this. Whereas, an article by business writer and author Kirk Kardashian entitled *When Retailers Do Good, Are Consumers More Loyal?* showed that when CSR activities are broken down, the results were more diverse.

When discussing a popular high-street supermarket, Kardashian categorised the main types of CSR activities possible as: environmental friendliness, community support, selling local products and treating workers fairly. Whilst the act of selling local produce had a strong universal appeal, it was discovered that the environmental campaign actually received a lot of negative responses from consumers – many of whom believed that the organisation should, instead, be using those resources to improve customer service.

In 'The Impact of Perceived Corporate Social Responsibility on Consumer Behaviour', Karen L. Becker-Olsen et al. argue that the use of CSR initiatives to positively influence consumers and effectively advertise products and services has now become commonplace.

Essentially, it has become customary for organisations for to use CSR initiatives to give their products and services a competitive edge – however, according to Becker-Olsen's research, these 'embedded promotions' don't always have a positive effect. In fact, low-fit initiatives have a negative impact on consumer beliefs, attitudes and intentions – regardless of the organisation's motivation.

It seems that initiatives that are not perceived to be aligned with the rest of the organisation's perceived goals, will been seen a negative impact. Moreover, any initiatives that are perceived to be high-fit but also profit-motivated will also have a negative impact.

Getting CSR right

Over the last few decades, thousands of organisations have pursued goals relating to corporate social responsibility with a broad – and, occasionally, even quite vague – idea of promoting increased wellbeing to the communities and social groups on which they depend.

However, more recently, as the concept has become more mainstream and started to be the stuff of much good marketing collateral, it has also become increasingly officialised – with many organisations even taking it on as formal 'business discipline' – and demanding that any initiative created in its name must deliver 'results', whatever they might be.

It seems like the spirit of what CSR once was has been lost somewhere along the line. Surely, when stripped back to basics, the original goal was simply to help align an organisation's social and environmental activities with its business purpose and values.

If, as a practical upshot, it also improved the organisation's reputation, helped mitigate risks, improved worker morale and improved performance, that's great – but, in the past, these things were secondary. Though they were obviously positive outcomes, they were never factored-in results. When it started out, before it came under the influence of PR and marketing, and key-performance indictors attached to it, CSR was something was simple. It might even be said 'pure'.

Over the course of the four-year period ending in 2015 *Harvard Business Review* (HBR) surveyed the business leaders that attended its CSR Executive Education Program, and asked a succession of questions about the range, organisation and management of their organisation's CSR campaigns and activities.

Though the HBR team was aware that, by attending the course, it meant that their sample was more likely to be composed of organisations with relatively advanced CSR agendas, it still found that around 60 per cent of the respondents said they were dissatisfied with their organisation's CSR activities and general approach to CSR and wanted to improve them.

The HBR study found that most of the organisations had fairly disparate views of CSR – ranging from pursuing 'shared value' (generating economic value in such a way that also creates value for society), to chasing pure philanthropy, to looking at environmental sustainability as the main target. In some cases, a mix of these goals was also attempted. In some cases, the motivation seemed to be based on improving brand image and public relations.

However, despite these broad visions of CSR, the team at HBR found that nearly all the organisations attending their course were hampered by an absence of coordination and lack of vision when it came to connecting their CSR programmes with the rest of their organisation's operations.

The HBR research also showed that though CSR is often lauded as something of tremendous importance to an organisation from its boardroom, most of the organisations' leadership were very hands-off. Usually, programmes are run – often in a very casual and uncoordinated way – by middle managers. Though leaders made a strong case for CSR almost universally, few were actively involved in operations.

Those few organisations that seemed passionate about making a positive impact on the environment and the communities they operate in, rather than demonstrating lip-service or creating campaigns simply for public relations opportunities, assigned resource to develop coherent CSR strategies – and directed this from the boardroom.

Research by the HBR shows that, regardless of industry and geographic location, organisations tend to focus the CSR activities upon the three areas of activities: philanthropy, improving operational effectiveness or transforming the business model.

- **Philanthropy** – this involves programmes where the main goal is not to create profits or in any other way improve the performance of the core business, but rather benefit the greater good. Examples of this might include charitable donations of money or equipment, promoting engagement with community initiatives or volunteer programmes. Technology giant Google is well-known for its philanthropic ventures and has various volunteer grants programmes, which provide funds to non-profit organisations – such as charities and youth groups. For every five hours that a Google worker volunteers, the company pays the non-profit organisation $50.

- **Improving operational effectiveness** – these sorts of programmes operate within existing business processes to deliver social or environmental benefits in ways that underpin the organisation's wider operations. They can also increase revenue. Examples include sustainability initiatives that reduce resource use, cut waste or lower emissions – which might, in turn, reduce business overheads. This may also involve investments in workplace equipment and conditions, health care or training – which may enhance productivity, retention and company reputation.
- **Transforming the business model** – this is where there are not additional programmes, but new areas of business adopted specifically to address social or environmental challenges. A frequent requirement of these sorts of programmes is to achieving community or environmental goals. For example, the global communications service provider BT has spent a lot of time over the last few years repositioning themselves as a sustainable and responsible business. They currently run the 'SOS Children's Villages' initiative which gives young people in isolated African communities the chance to improve the quality of their lives by learning digital skills. At the same time this action increases the organisation's network and provides a foothold into untapped market areas.

Creating cohesion

Even if leaders are prepared to get behind their organisation's CSR activities and have successfully mapped out what they want to do, getting started can still be a daunting task.

Many CSR programmes are poorly coordinated. Before looking at getting new plans in place, organisations need to look at their current portfolio (if they have one) and remove any initiatives that don't promote social or environmental issues – and, crucially, that don't align clearly with the organisation's purposes or values. For example, it would make more sense for a clothing supplier to offer its workers operating in Bangladesh free baby clothes for their children – and lighten their burden that way – than it would for the clothing supplier to offer them a team-building away-day to a zoo. Aligning is about ensuring that the activities chosen for CSR are consistent with the organisation's business purpose – and that they have a valued outcome.

Organisations also need to define what success looks like. As we've seen, not all initiatives are designed to generate revenue, some are designed to cut expense, and some are created to improve performance. So, working out what goals are and how they might be recorded or measured is an essential first step.

Naturally, CSR initiatives will be – or, at least, *should* be – different between organisations. If they have successfully created coherent campaigns that are consistent with their values, this obviously means that they need choose something personal to their own industry or sector.

Most initiatives based on improving operational effectiveness measure success by way of tangible methods, such as working out how much energy has been saved or how much waste cut. After some wilderness years – in which they soaked up a lot of bad press – at the start of 2007 search giant Google suddenly started made aggressive moves towards self-improvement – and it's clear that the effort is paying off.

The organisation's initiative called 'Google Green' has been a considerable success in using resources more efficiently and supporting renewable power. This extends to much more than just buying recycling bins and installing low-energy lightbulbs, Google's investments have been radical and wide-ranging – including substantial investments in wind, solar, solar thermal and geothermal technologies – which have already contributed to a remarkable 50 per cent drop in power requirements at many of their data centres.

Instead of investing in technology, the American package delivery company, UPS, regularly employs independent auditing companies to assess where it can reduce energy use and carbon emissions – and, as a result, the company now publishes impressive cost and resource savings on an annual basis.

One organisation that might be said to epitomise 'shared value' method is the American fashion brand TOMS. For every customer that buys a pair of TOMS shoes bought in one of their stores, a pair of shoes is gifted to a charitable organisation in the developing world. When customers buy their branded sunglasses, money is given to medical charities specialising in giving disadvantaged individuals eye-restoration treatments. The company's business model proves that customers are willing to pay a premium for a good cause – and shows that it can be very powerful when clearly aligned with the organisation's operational endeavours.

Interdisciplinary CSR

Instead of relying on a number of middle-managers to design and implement CSR strategies in addition to their normal workloads, which – according to HBR's research – is often the case, organisations need to display coordinated support for CSR initiatives from the top level if they wish them to succeed.

If an organisation is to take CSR seriously, they should establish a role for an individual whose primary responsibility is leading CSR. This is even more important if they are attempting interdisciplinary initiatives that integrate the three types of CSR initiatives: philanthropy, improving operational effectiveness or transforming the business model.

This level of coordination is rare, though, and the range of purposes underlying initiatives can often cause confusion and major management issues, even if a dedicated member of staff is assigned to oversee it. Why is this? The problem is that the three desperate types of initiative tend to stem from very different parts of the business. Philanthropic programmes will often be under the purview of Community Managers and typically be affiliated to the organisation's HR Department. Whereas, initiatives based on improving operational effectiveness are often run by Operations Managers and usually report to the Head of Sustainability, Chief Sustainability Officer or similar. When it comes to shared-value initiatives, CEOs are sometimes loosely involved, but the research of HBR indicates that only around 30 per cent of organisations have any involvement from a senior executive. In a small number of cases, CEOs might have some operational influence – but in the vast majority of cases, these sorts of initiatives are also handed over to a lower-level worker to manage.

It is little surprise, therefore, that organisations often struggle to create a coherent vision for their interdisciplinary initiatives, when the usual state of CSR endeavours is that responsibilities are spread across the organisations and among several individuals at a range of levels.

The Harvard Business School advocates two effective approaches to CSR strategy development. Let's look at these now through two examples.

- **Top-down** – Swedish flat-pack furniture giant IKEA has used the top-down method to CSR. All of the organisation's strategies and visions are down to a small executive management group, made up of just seven senior managers. For 2011 onwards, the company has pursued plans that combine aggressive market growth and sweeping sustainability plans. It also devised a social welfare initiative known as 'the IKEA Way' – a succession of programmes focused on maintaining labour standards throughout their complete supply chain, and, in particular, preventing child labour. The organisation's sustainability agenda is not only seen as morally virtuous, but has the additional weight of being developed – and repeatedly extolled – by organisational leaders, thus imbuing it with considerable authority.
- **Bottom up** – This less conventional approach has seen some notable successes. For example, in 2010, Ambuja, India's top cement supply company, set up The Ambuja Cement Foundation – a new CSR wing of the organisation dedicated to taking charge of all social and environmental initiatives. It immediately instigated committee meetings at a plant-level with a view of escalating issues of organisation-wide importance to the boardroom. Any issues flagged, were then subsequently discussed at the boardroom level. Since its inception, The Foundation has devised an impressive range of philanthropic projects including building a number of integrated rural development projects. They have also cut energy and pollution levels across their plants. And the business has diversified into a number of different areas including: community health and sanitation, education and female empowerment. The ambitious CSR targets have also seen the organisation making a commitment to improving their CSR output year-on-year – and giving back more than it takes from the local community.

Since organisations are influenced in countless ways by all manner of factors – such as industry, location, age and societal and community issues – it makes little sense for them to engage in the same types of CSR activities. After all, there is no such thing as an off-the-shelf 'right' choice for CSR – in order to succeed they need to be personal, appropriate to the nature of the organisation and mirror the motivations of the workers and the people that govern the organisation.

For example, a clothing manufacturing business might have considerable scope to tackle a variety of issues within its supply chain – that an internet services company is less likely to need to address. However, the latter might be well placed to use its social channels to affect a great deal of positive change in offering free online advertising for charities and help groups. On the other hand, for organisations operating in locations that enjoy strong provisions for social welfare, the CSR initiatives might look entirely different to those working out of countries where there is little sufficient government funding for public health.

THE NAKED TRUTH: CORPORATE SOCIAL RESPONSIBILITY

As we explore the breadth and depth of CSR as an organisational discipline, the parallels with OSH quickly become apparent. The three elements identified in the HBR research: *philanthropy*; *improving operational effectiveness*; and *transforming the business model* could easily be pillars of an organisation's OSH agenda. And Harvard's two effective approaches to improving CSR: *bottom up* and *top down* are certainly often used when it comes to roll-out of workplace safety initiatives.

Now consider those four interconnected responsibilities of Carroll's *Pyramid of Corporate Social Responsibility* (economic, legal, ethical and philanthropic) and it's abundantly clear that the similarities between the two disciplines are many. CSR, just like OSH, is all about working for the greater good – of the organisation, its people, its stakeholders and its local communities.

Yet in CSR as in OSH the value of the discipline is often overlooked – until it's too late and an adverse event occurs, forcing organisations into the limelight and under scrutiny.

In an ever-more connected world, CSR – like OSH, provides myriad opportunities for organisations not simply to do the right thing, but to create deep value for its people and the communities in which it operates, providing stability, sustainability and success for all.

GET NAKED: CORPORATE SOCIAL RESPONSIBILITY

1 How does your organisation consider its social responsibilities? Look back the HBR types of CSR initiatives (philanthropy, improving operational effectiveness, transforming the business model) – does your organisation take a balanced approach across these three types or tend to focus in one area?

How do you feel about this? Where are the opportunities to leverage what you do across the other types of initiatives?

2 Would you describe your organisation's strategic approach to CSR as mainly *top down* or *bottom up*? What impact does this approach have on your key stakeholders both within the organisation and in the broader community? What are the pros and cons of each approach?

3 Are all four elements of the *Pyramid of Corporate Social Responsibility* (economic, legal, ethical and philanthropic) included in this approach? How does each of these align with local need from the societies in which your business operates?

4 How is responsibility for CSR allocated in your business? Is there a single person tasked with managing the three types of CSR initiatives (philanthropy; improving operational effectiveness; transforming the business model)?

5 Often creative ideas like the ones discussed at TOMS, Ambuja and Google come from the workforce, rather than senior management. How does your organisation encourage employees to contribute to the CSR agenda?

6 Where are the opportunities to align or link CSR activities with the OSH approach in your organisation?

19 The future of safety

Even in those regions of the globe where Occupational Safety and Health (OSH) legislation is robust and comprehensive, safety practitioners still face an uphill battle.

As the world progresses, organisations quickly follow, acclimatise and adapt. New research, fluctuating social trends, political and economic changes and advances in technology, all have the capacity to create new workplace dangers – and the role of every safety practitioner is complicated by the fact that they are attempting to hit what is essentially a moving target.

In this chapter, we'll look at some of the current trends in health and safety, consider why looking back may help us move forward and explore what organisational safety might look like in the future.

Despite considerable resources being thrust at safety management programmes, the improvements in health and safety statistics seen over recent years – which saw huge falls in the number of workplace deaths, illnesses and accidents – have started to trail off of late, and accident numbers have begun to plateau.

Whilst organisations investigate the potential reasons for this, the true issue might simply be one of definition. After all, does a lack of accidents really equate to 'safety success'? It might seem like the obvious characterisation – but, at the same time, that would seem to suggest that *all* organisations are a safety success *until* they have an accident. So, really, it is not a useful description.

Forward-focused safety practitioners and business leaders around the world realise this and are beginning to see safety as something that needs to be calculated through successes rather than failures – and, thus, instead of seeing safety as something that is 'achieved' – it is now seen as something to be 'created'.

In occupational safety, looking towards the future is hugely important. After all, organisations are constantly evolving – altering course, adjusting strategies, adapting missions, changing priorities, amending processes, improving technology and so on. Consequently, our approach to safety needs to evolve with them in order to remain effective.

Whilst it is impossible to create a perfectly accurate prediction of what safety practices might look like in the future, it is possible to view the current trends in the safety sphere and draw some inferences as where things might be heading.

So, in this chapter we'll consider some of the more prevalent industry trends, and strip these back to gain insight into current priorities – and, by extension, the shape of what might be coming your way.

Integrated systems approach

The coming decade will see many more organisations integrating their safety processes and management systems more fully into their operational activities and structures – which will naturally result in safety becoming embedded into all the organisation's operations.

Research shows the journey to an accident-free workplace and sustainable operational growth both share their roots in a solid foundation in safety. The only way to ensure this is achieved is to integrate safety within a wider organisational context, so that it rests within a holistic operational risk and management strategy designed to grow sustainable performance improvement.

The new international standard ISO 45001, whilst designed to provide a framework approach for occupational safety and health management systems, will certainly facilitate this integration. The new standard has been constructed in order to provide easy linkage to other systems including approaches to quality (ISO 9000) and environmental (ISO 14000) and risk (ISO 31000) management. ISO 45001 is centred on a model of *Plan, Do, Check, Act* – already familiar to many working in the world of occupational health and safety. It utilises the same high-level structure (common clauses, terms and definitions) as all new management system standards to enable ease of integration. Furthermore, at least

40 per cent of the core text of all new management system standards will be standardised, making it not just easier to integrate different disciplinary approaches, but also make it easier to conduct audits and governance activities.

Focus on leading indicators

In the last year or so, there has been a realisation that counting the number of times organisations have failed at safety and reporting this as an accident rate is neither meaningful at measuring safety nor effective at motivating action. Organisations will move away from lagging indicators to concentrate on leading indicators. This shift will mean that organisations need an attitudinal shift from a position of 'doing less harm' to one of 'doing more good'.

Whilst this might seem – on the surface, at least – like a commendable, more constructive, approach, there remains a risk that the motivating factors behind such a move could be fundamentally about protecting the organisations' reputation, maintaining their licence to operate or simply promoting a brand.

This sort of shallow commitment to safety equates to little more than good public relations. The result, in fact, is that much less is being done to leverage safety and sustainability – resulting in a slump in efficiency, innovation and competitive advantage.

To help make the shift, a balanced scorecard approach – retaining the traditional injury and accident rates whilst transitioning to measurement of positive inputs – will be adopted and rather than gauging success on a lack of adverse events, organisations will develop and utilise a 'safety index' comprising several measures, tied to specific actions and weighted to reflect potential impact of activities. These balanced scorecards and indices will help organisations to properly monitor their management systems and drive continuous improvement.

Human capital risk management

As well as assessing workplace risks and implementing hazard controls, the role of the safety practitioner will evolve and expand to take responsibility and shape influence over a range of broader people-related issues. Right now, in developed nations around the world the topics of health and wellbeing of workers are already falling under the remit of the safety manager. As a result, practitioners in the future will need to extend their range of core skills.

But health and wellbeing are not the only areas in which safety professionals will be increasingly asked to lead in. In the future, a push towards 'human capital risk management' will prevail with organisations ring-fencing people-related risks such as ethical business practices, security, travel risk management, corporate social responsibility, reputational risk and more and aligning these with health, safety and environmental operational jurisdictions.

As more holistic approaches to and definitions of safety are adopted over the coming years, it will mean health and safety professionals will be required to broaden their range of skills and competencies. Rather than eschewing these subjects as things outside their remit, and concentrating on their perceived 'core roles', the more progressive practitioners will recognise the value and impact these factors have on organisational health and safety – and choose to 'upskill' themselves and embrace these new disciplines. If they do not, they are liable to fall behind.

Workers as the solution

Workers have traditionally been as one of the 'risks' within organisations – just another factor that needs to be controlled to create safety in the workplace.

This control was typically asserted by limiting their actions or by placing constraints between them and how they performed operations. However, many organisations are now beginning to undergo something of a paradigm shift – and are beginning to adopt a more 'people-positive' approach, originally advocated by renowned psychologist Sidney Dekker.

In *Safety Differently: Human Factors for a New Era*, Dekker suggests that organisations need to listen to workers in that they: "Encourage and support them because they are as interested in the health, success, longevity and profitability of the company as is any executive."

Dekker describes how organisations should go about transforming safety from simply bureaucratic liability into an ethical responsibility, and outlines how human workers should not be seen as a problem to control – but as a 'solution to harness'.

Rather than having line managers and compliance auditors controlling workers more progressive organisations will begin adopting systems where safety is defined as the pursuit of the positive goal of ensuring things 'go right' – and where direction is given by whoever is best able to do so. Fundamentally, this approach seeks to ensure that workers are not guided by oppressive rules and regulation – but by insight and context. This makes great sense.

Looking to changes in management systems, it's worth noting that ISO 45001 auditors can now go to any level of the organisation during their interventions – 'worker participation' has been defined and added to the international standard within the leadership section.

Focus on trust and engagement

In striving to improve safety, organisations will spend more effort conducting competitor research and analysis of top-performing organisations in order to find out what factors are important to organisational safety. Invariably, they will discover that leadership is the major contributing factor.

It is not by chance that the best safety performers are also those organisations that have strong leadership. But just what is it that makes leadership 'strong'?

One thing: trust.

Trust in its leaders is what holds strong organisations together – and leaders that command trust have the ability to inspire workers and positively influence behaviour in a much more effective way than any regulation or compliance.

Over the course of the last few decades, organisations have started to realise that there is little point in having trusted and influential leaders if their activities are entirely confined to the boardroom. It has become clear that leaders need to ensure that trust is established – and spreads virally throughout the workplace – by engaging with workers.

As organisations continue to become more laterally structured, it means that traditional boundaries will be broken down, silos dissolve and different parts of the organisation will start to work more effectively together – directed, of course, by trusted leaders clearly articulating the mission.

By engaging with workers and communicating across all levels, not only will this establish an inclusive atmosphere, but it will remove any chance of a perceived 'them and us' ethos developing. It also lessens the risk of the phenomenon of 'groupthink' emerging – which often leads to irrational decision-making.

With the trend in organisational structure moving away from strict demarcation, and previously rigid job roles becoming freer to reflect the greater need for knowledge-sharing, the traditional hierarchical control mechanisms will naturally fall away – and, in the future, organisational success will become dependent on workers trusting their leaders.

Deeper cultural understanding

Understanding exactly just how things get done is essential knowledge for any forward-thinking organisation – but it is not always that easy. Culture – as an assembly of values, beliefs, attitudes, behaviours, systems, leadership and more – is dynamic and constantly in a state of flux. They shift, subtly, in all manner of ways, influenced by a wide variety of internal and external factors.

As globalisation continues, joining together countries, continents and cultures, moving forward organisations will invest more time and effort to better understand their corporate culture and highlight the factors that either enable or restrict progress.

Cultural assessments that provide deep insight into how people think and what they do at work will form part of the ongoing process of business intelligence. Safety climate surveys comprising myriad questions for employees to check their level of agreement will be recognised for their ability to provide a snapshot of average thinking and will be eschewed in favour of assessments that use real people (rather than machines) to engage with workers to seek opinions, ideas and test hypotheses.

This shift will not only help safety professionals and business leaders understand what elements affect culture but will naturally reinforce that culture change *can* be both managed effectively and measured efficiently.

New values

As working conditions have developed over the last generation, so has the nature of the relationship between workers and employers.

The informal 'psychological contract' – what workers and employers expect from one another – has radically altered over the last 30 years. Yet, few organisations seem to recognise that whilst what they offer has changed, there has also been a corresponding change in how workers perceive their conditions of work.

Workers no longer expect – or, in fact, want – lifelong employment within a single organisation. The old 'contract' that their parents or grandparents might have entered into – based on loyalty, job security and steady advancement – is now largely confined to history.

Though career advancement is still a motivating factor, most modern workers neither expect job security or loyalty to be part of the deal. The new contract seems to be mainly that of an expectation of continued learning and development – which has presumably developed to fill the hole left behind in the absence of job security, as knowledge and skills are transferable between roles and seen as an enabler for advancement.

In *The Support Economy* James Maxmin and Shoshana Zuboff state that the lack of job security in the modern marketplace has created a new individualism amongst workers – who are now invested in 'psychological self-determination'. This has ultimately led to feelings of 'corporate indifference'.

As well as expressing a need for ongoing personal development, modern workers also want participation, expression, satisfaction and quality of life. Whilst, these are all ideals often espoused by organisations as 'values' – too few pro-actively promote or encourage them. As a result, workers are often left apathetic, and have very little to motivate them.

Progressive companies will be those that understand that traditional drivers are largely redundant in the unstable, fast-paced modern marketplace, where many workers have been conditioned to feel temporary even in permanent roles. Fostering a sense of purpose, through delivering meaningful work that is recognised by business leaders, will be the focus for organisations at the front edge.

Causes of deviations

Based on Walter A. Shewhart's 1931 research paper *Economic Control of Quality of Manufactured Product* – and, therefore, not exactly a new idea – this approach will become fashionable again over the course of the next few years.

As most workplace incidents are caused by deviations – or weaknesses – in the way the organisation operates, Shewhart's methodology will help by bringing focus to identifying and removing any flaws from the day-to-day processes.

The shift in focus from 'causes of incidents' to 'causes of deviations' will mean that many organisations can move towards making positive changes to their operations, rather than simply focusing on eliminating the causes of incidents.

Safety in numbers

So-called 'big data' has taken safety by storm over the last few years – and, accordingly, safety practitioners are finding it necessary to gather more data than ever before.

Big data is made up of vast data sets that can be analysed computationally to reveal patterns, trends and associations. The analysis of safety data is likely to become incredibly important – particularly within the field of Behaviour-Based Safety (BBS) as it has the potential to provide powerful insights into human behaviour.

Currently, 'big data analysis' is problematic, since the volume and complexity of the data makes it difficult for humans to assess and appraise – and, consequently, errors in the algorithms are liable to go undetected.

Discovering a method to manage this vast data in a way that is organised, efficient and easy to assimilate will be a huge step forward for safety. Smart organisations will recognise the opportunity that big data brings, without falling into the trap of multiple measurement for measurement sake. Balancing the scorecard with measures that reflect what the business values will be key.

Creating safety

Generally, organisations have approached the challenge of workplace safety by working out what might cause injury and then implementing controls to reduce the risk of harm occurring. Typically, following an accident many organisations leap into action with formal investigations to identify the causes of the adverse event, and then set about rewriting policies and procedures, retraining staff and issuing edicts that remind everyone to be safe at work. When these actions appear to prevent an accident recurring, success is deemed to have taken place.

But the absence of accidents is not equal to the existence of safety. Counting the number of days since the last accident tells us nothing, really. Myriad factors – including luck and under-reporting – could be the real reason behind a drop in accident rates.

But let's strip things back further. If you think about it, rushing around trying to prevent recurrence of workplace accidents is fundamentally flawed. The approach relies on having an accident to act as the catalyst for action.

When asked about his excellence in the sport of boxing, Cassius Clay (perhaps better known as Muhammad Ali) advised that his secret was not to count the days he holds a championship title, but rather to make the days count. Intrigued, the reporter duly noted down the legend's words as he explained how he concentrated on the *inputs*: his diet, training regime, strategy to choose opponents and tactical advice from his coach when in the ring. This, Ali shared, seemed to bring about the outcome he desired: winning the match. The boxer's committed forward-focus brought him significant success including retaining the record for most heavyweight fights won which was unbeaten for more than 35 years. Furthermore, Ali was ranked as the greatest athlete of the twentieth century by *Sports Illustrated* magazine and Sports Personality of the Century by the BBC.

So, what can we learn from the man nick-named *The Greatest*? Well, in the future the focus for improvement will shift to become more on the inputs to creating a safe work environment. A mind-set that seeks to accentuate the positive rather than strive to eliminate the negative will be the foundation for action as organisations set out to manage (and measure) the activities that really add value. These will include (but are not limited to) team talks, shared best practice learnings, lean six sigma projects, employee engagement and much more.

Creating safety (see next chapter) rather than preventing accidents will bring the clarity of commitment needed to slice through the mystery and negative stigma that has attached to workplace safety and drive positive engagement, encourage action, empower people and enable success. Now we're really talking about an evolution!

THE NAKED TRUTH: THE FUTURE OF SAFETY

With technological advancement and regulatory change transforming the corporate landscape faster than any time previous in history, there has never been greater need to have a clear view of risk embedded into every aspect of organisational life.

Workplace safety is the by-product of the decisions made by leaders, the adopted mission, the corporate strategy, the prevalent culture and the performance of workers.

Moving forward we will see an upsurge in the number of organisations recognising the fact that safety needs to stop being viewed as an independent process, disconnected from the rest of day-to-day operations and, instead, should be seen for what it is: the outcome of those operations.

With greater understanding, safety practitioners and business leaders will work hard to present and facilitate safety not as an operational add-on but as something integrated into every element of business thinking and behaviour.

A concerted mind-set shift will occur, where forward-looking practitioners and leaders successfully convince organisations and industry sectors that efforts must be realigned to the input activities that generate the outcome of a safe workplace. These people will focus on creating safety rather than preventing accidents.

Leadership remains vital and will be the primary route to achieving and sustaining safety excellence. Strong safety leaders will enable excellence by engaging fully with workers, by setting clear expectations, encouraging collaboration in pursuit of a shared vision, role modelling great behaviours and empowering everyone at work to take action.

In the future, this – more than anything else – will be critical to safety success.

GET NAKED: THE FUTURE OF SAFETY

1. Describe the ideals that drove the founding of your current organisation. Do you think they are still relevant? Are they sustainable? Will they help the organisation to flourish in the future?
2. What will an integrated systems approach mean for you as a business leader or safety practitioner? What do you need to do now to ensure that you can leverage success for your organisation with an integrated approach?
3. Consider the way that your organisation considers safety performance right now. Do the metrics suggest that the organisation measures what it values or values what it measures?
4. What are the human capital risks you might anticipate for your organisation in the future? Where are the opportunities for you to develop your skills now in advance of these changes?
5. How successfully does your current organisation adapt to change? Think about changes that have happened from inside *and* outside the organisation? What should the organisation focus on and pay attention to?
6. How engaged are your employees in the safety agenda right now? Does their contribution build meaning and generate satisfaction for those involved? How could you engage more of the workforce in developing safety solutions?
7. Is your organisation looking backwards at trying to prevent accidents or focused forward on creating safety? What ideas do you have to shift the mind-set towards the inputs required in order to achieve the outcome of a safe workplace?

20 Creating safety

Workplace safety has long been the subject of discussion in organisations around the world, from the boardroom to the shop-floor. Yet today, in a growing climate of risk aversion and hazard elimination fuelled by negative media hype and over-zealous law firms, the answers to contemporary safety challenges cannot be found in the legislation, nor in the creation of more policy or procedure.

Making positive safety changes within an organisation is often approached with anxiety or apprehension, so the aim of *Naked Safety* has been to emphasise that it can be a straightforward process when we strip things back to what really matters.

And when we do that we realise that at the heart of workplace safety is one thing: people.

Through the chapters in this book, such topics as beliefs, values, behaviours, communication, engagement and leadership have been explored. Each of these feeds directly into an organisation's culture and helps to shape and sustain it. Therefore, for anyone wishing to make a positive impact upon safety culture, it is essential to recognise these factors and understand how they interplay.

Where organisations typically fail in making inroads towards safety performance improvement is that they are often more *reactive* than *proactive* in their perspective, habitually dwelling on past incidents and trying to prevent recurrence rather than looking to the future to effect positive change. Whilst learning from accidents is arguably useful, this approach is counter-productive on its own, literally causing businesses to sit and wait for the next bad thing to occur. To implement successful safety programmes and have a positive impact on safety culture organisations need to begin with a change of mindset.

It is not by coincidence that the world's most successful organisations are the ones where safety underscores every single process. Within these organisations, safety has ceased to be about one single individual 'policing' safety: it has developed into something that is considered and maintained by each and every worker, irrespective of level and role.

Of course, when it comes to worker safety, there is no 'silver bullet', no magic wand, no 'one-size-fits-all'. Every approach will always need to be fine-tuned to incorporate specifics based on industry, geographical location and other factors. Though when it comes to truly world-class safety systems, there's a lot more in common than that which divides them. Let's strip these back now as a way to round off our journey.

Forward focus

Organisations that take a forward focus concentrate on the inputs to creating a safe work environment and use leading indicators to enable them to identify and manage potential problems *before* they arise.

Whilst it is important – and, in many states, a legal requirement – to record and report injuries, when it comes to workplace accidents past performance is rarely a strong predictor of future outcome. This point is well illustrated when we consider the Deepwater Horizon oil rig, which exploded in the Gulf of Mexico in 2010, killing 11 people and creating a gamut of long-term environmental problems. It later transpired, that the day before the accident, senior leaders from Transocean, the rig's owners, had visited the site to celebrate several years without a significant safety issue. The message was clear: the absence of accidents does not equal the existence of safety.

Evidence-based decision-making

All too often for many organisations safety improvement relies on guesswork or instinct, rather than real data.

This is dangerous, as even the most competent safety practitioner can become blindsided to the whereabouts of potential threats. Following accidents, it is not unusual for experienced safety people to be left scratching their heads wondering just what happened. After all, experience does not guard against the unknown.

Good safety relies on adopting a measured approach based on carefully collating and analysing information from a wide range of sources. Only when this happens, is truly informed decision-making possible.

Communication flow

Good communication within an organisation is key to great safety performance. It keeps workers up-to-speed with new and existing policies, procedures, task requirements and with lessons learned – both from adverse events such as workplace accidents and near misses and also in the sharing of best practice.

Crucially, it creates avenues from the front-line staff to upper management for consideration in the development of organisational strategies, focus and day-to-day approach. This flow of information – in both directions – is an essential element in creating safety in any organisation.

The importance of opportunity

It's a fact that near misses are greatly under-reported. And for many organisations they're greatly under-valued too.

Workers typically – and, perhaps, not unreasonably in some organisations – assume they will be chastised if they admit to being involved in a near accident or near injury. However, near misses are vitally important early warnings and need to be embraced fully.

Organisations need to encourage the reporting of all near misses that workers experience and any stigma around such an action needs to be quickly removed. Indeed, many organisations are now introducing anonymous reporting systems that focus on information gathering rather than apportioning blame so that the opportunities for learning from near misses is maximised.

Meaningful training

For many organisations training has developed into something imposed on workers. Quite often it takes place in an over-formalised setting, where workers feel uncomfortable, and have little understanding as to why it's happening. Workers arrive, sit down, sign in and nod sagely at all the right moments, before traipsing back off to work.

The training might be essential – learning a new safety system, for example, or following the introduction of a new legal directive – but where there is little meaningful explanation to the workers actually being trained, it always results in a marked lack of enthusiasm.

Retention of information rarely occurs when subjects are not 'switched on' or in an open, positive state of mind. Current research indicates that around 85 per cent of what we learn comes from observation of our peers, and as little as 10 per cent from formal training.

Human beings are endowed with natural curiosity – and, in environments where they feel comfortable, will usually find picking up new skills a pleasurable experience. How inclusive is your approach to safety training?

A positive work environment

Research shows that a happy and engaged workforce is considerably more productive. Conversely negative working environments have an immediate and lasting financial impact through the creation of disengaged workers, high staff turn-over and absenteeism.

Cheerfulness has been found to be one of the major factors in maintaining safety at work. Creating a cheerful working environment is not difficult, it's just something that most leaders say they have little time to do. But this makes no sense: leaders at every level make time to give orders, instruct action, provide direction. Rather than taking a 'telling' approach, if leaders were to adopt a more facilitative style this would certainly boost morale and positively impact job satisfaction too. Interestingly, workers with low job satisfaction are three times more likely to be injured at work than those with high job satisfaction. Accidents and injuries are also far more likely to occur in high-stress work environments.

Creating safety

The overarching theme behind *Naked Safety* is that within any organisation, safety leadership shouldn't be a hierarchical duty or some lofty ideal but rather something that at a much more fundamental level is woven into every worker's role, regardless of their place within the corporate structure.

Improving safety culture isn't the exclusive preserve of the safety professional. Organisations that think like this, or employ precious or reclusive safety teams, need to quickly update their thinking. The onus should never be on a single individual or department 'taking care' of safety – instead, it needs to be about making it a value shared by the entire organisation from top to bottom.

Creating safety isn't a difficult or onerous task either. The first step forward is that rather than adding extra layers of compliance or bringing in complicated safety programmes organisations should find ways to actively encourage every member of their workforce to be mindfully proactive with regards to safety. This is not the same as telling them to 'be more careful'.

Naked Safety is about creating an inclusive, engaging and pragmatic approach to managing risks – and, in doing so, building positive changes that empowers people to do the right thing, and generating significant and sustainable positive impact in an organisation where everyone goes home without harm at the end of every day. Go on, strip it back.

By the same author

The Wellbeing Book (2018) LID Publishing.

Working Well (2018) Maverick Eagle Press.

Safety & Health for Business (2017) IOSH Publications.

From Accidents to Zero: A practical guide to improving your workplace safety culture Second Edition (2016) Routledge.

Mind Your Own Business: What your MBA should have taught you about workplace health and safety (2016) Maverick Eagle Press.

Safety Savvy (2015) Maverick Eagle Press.

From Accidents to Zero: A practical guide to improving your workplace safety culture (2014) Maverick Eagle Press.

Bibliography

Adair, J. (1989) *Great Leaders*. London: Talbot Adair.

Adair, J. (2010) *Strategic Leadership*. London: Kogan Page.

Adams, R., Owen, C., Scott, C. and Parsons, D.P. (2017) *Beyond Command and Control: Leadership, Culture and Risk*. Boca Raton, FL: CRC Press.

Advisory Committee on the Safety of Nuclear Installations. Health and Safety Commission (1993) *ACSNI Human Factors Study Group. Third Report. Organising for Safety*. HMSO. Research Report 36.

Allan, N. and Beer, L. (2006) *Strategic Risk: It's All in Your Head*. University of Bath, UK. Working Paper Series: 2006.01.

Antonsen, S. (2017) *Safety Culture: Theory, Method and Improvement*. Boca Raton, FL: CRC Press.

Aristotle (1953) *The Ethics of Aristotle: The Nichomachean Ethics*, trans. J.A.K. Tomson. London: Allen & Unwin.

Armstrong, M. (2006) *A Handbook of Human Resource Management Practice*. London/Philadelphia, PA: Kogan Page.

Babyak, M., Blumenthal, J.A., Herman, S., Khatri, P., Doraiswamy, M. . . . and Krishnan, K.R. (2000) Exercise Treatment for Major Depression: Maintenance of Therapeutic Benefit at 10 Months. *Psychosomatic Medicine*, 62(5), 633–638.

Banerjee, S.B. (2004) 'Reinventing colonialism: Exploring the myth of sustainable development', *Situation Analysis*, 4 (Autumn), 95–110.

Becker-Olsen, K., Cudmore, B.C. and Hill, R. (2005) The impact of perceived corporate social responsibility on consumer behavior. *Journal of Business Research*.

Beer M., Spector B.A., Spector B. (1990). *The Critical Path to Corporate Renewal*. Boston, MA: Harvard Business School Press.

Bennett, T., Grossberg, L. and Meaghan, M. (eds) (2005) *New Keywords: A Revised Vocabulary of Culture and Society*. Oxford, UK: Blackwell.

Berle, A.A. and Means, G. (1932) *The Modern Corporation and Private Property*. New York: Macmillan.

Bernardi, L., Porta, C., Casucci, G., Balsamo, R., Bernardi, N.F., Fogari, R. and Sleight, P. (2009) Dynamic Interactions between Musical, Cardiovascular, and Cerebral Rhythms in Humans. *Circulation*, 119, 3171–3180.

Blanchard, K. and Johnson, S. (2003) *The One Minute Manager*. New York: William Morrow.

Blanchard, K. and Stoner, J.L. (2011) *Full Steam Ahead! Unleash the Power of Vision in Your Work and Your Life*. BK Business.

Boehm, J.K. and Lyubomirsky, S. (2008) Does Happiness Promote Career Success? *Journal of Career Assessment*, 16, 101.

Boholm, A. (2005) *Anthropology and Risk*. London: Earthscan.

Boyce, T., Robertson, R. and Dixon, A. (2008) *Commission and Behaviour Change: Kicking Bad Habits Final Report*. London: The Kings Fund.

Brandes, V. et al. (2008) XII World Congress of Music Therapy, Buenos Aires.

Bregman, P. (2011) *18 Minutes: Find your Focus. Master Distraction and Get the Right Things Done*. London: Orion.

Brokaw, L. (2009) *Does Sustainability Change the Talent Equation?* Fall 2009 Magazine: Opinion and Analysis. MIT Sloan Management Review.

Brooks, D. (2011) *The Social Animal*. London: Short Books.

Brown, S.L. and Eisenhardt, K. (1997) *The Art of Continuous Change: Linking Complexity Theory and Time-paced Evolution in Relentlessly Shifting Organizations*. McKinsey and Company.

Bryant, M.C. (1975) *Proceedings: A Safety Officer's View of the Future*. Medline, 18(4).

Buckingham, M. (1999) *First Break All the Rules*. London: Simon & Schuster.

Buttolph, M. (1999): Styles of Safety Practice: Monks, Mercenaries and Missionaries. *Safety & Health Practitioner*, March.

Cadbury, A. (2002) *Corporate Governance and Chairmanship: A Personal View*. Oxford, UK: Oxford University Press.

Callaghan, G. (2002) *Faith, Madness and Spontaneous Human Combustion*. New York: St Martins.

Carlowski, G. (1994) Anatomy of a Leader. *Management Review*, March, 12.

Carroll, A. (1991) The Pyramid of Corporate Social Responsibility: Toward Moral Management of Organizational Stakeholders. *Business Horizons*, July–August.

Chabris C. and Simons, D. (2011) *The Invisible Gorilla: And Other Ways Our Intuition Deceives Us*. New York: Harper Collins.

Chan, M.F, Chan, E.A, Mok, E. and Kwan Tse, F.Y. (2009) Effect of Music on Depression Levels and Physiological Responses in Community-Based Older Adults. *International Journal of Mental Health Nursing*, 18(4), 285–294.

CIPD (2003) *Managing Employee Careers – Issues, Trends and Prospects*. London: Chartered Institute of Personnel and Development.

Cohen, G.A. (2013). *Finding Oneself in the Other*. Oxford, UK: Princeton University Press.

Cooper, D. (2013) *The Strategic Safety Culture Roadmap*. Franklin, IN: BSMS.

Corporate Leadership Council (2004) *Driving Performance and Retention Through Employee Engagement*. Corporate Executive Board. 2004 Council Teleconference.

Costa, P.T. and McCrae, R.R. (1980) Influence of Extraversion and Neuroticism on Subjective Well-Being: Happy and Unhappy People. *Journal of Personality and Social Psychology*, 38, 668–678.

Covey, S. (1998) *The 7 Habits of Highly Effective People*. New York: St Martin's Griffin.

Cox, S. and Flin, R. (1998) Safety Culture: Philosopher's Stone or Man of Straw? *Work and Stress*, 12(3), 189–201.

Dawkins, R. (1999) *The God Delusion*. London: Transworld Digital.

Dekker, S. (2007) *Just Culture*. Farnham, UK: Ashgate.

Dekker, S. (2008) *The Field Guide to Understanding Human Error*. Farnham, UK: Ashgate.

Dekker, S. (2011) *Drift into Failure*. Farnham, UK: Ashgate.

Dekker, S. (2014) *Safety Differently: Human Factors for a New Era*. Farnham, UK: Ashgate.

DeMarco, T. and Lister, T. (1999) *Peopleware*. New York: Dorset House Publishing.

Deming, W.E. (1982) *Out of the Crisis*. Cambridge, UK: Cambridge University Press.

DeNeve, K.M. and Cooper, H. (1998) *The Happy Personality: A Meta-analysis of 137 Personality Traits and Subjective Well-being*. Waco, TX: Baylor University.

Diener, E. (2000). Subjective Well-being: The Science of Happiness, and a Proposal for a National Index. *American Psychologist*, 55, 34–43.

Douglas, M. and Wildavsky, A. (1982) *Risk and Culture*. Berkeley, CA: University of California Press.

Drucker, P. (2001) Taking Stock. *BizEd*, November–December, 13–17.

Drucker, P (2004) *What Makes an Effective Executive?* Boston, MA: Harvard Business Review Press.

Dweck, C.S. (2007) *Mindset: The New Psychology of Success*. New York: Ballantine Books.

Eccles, R.G., Newquist, S.C. and Schatz, R. (2007) Reputation and its Risks. *Harvard Business Review*, February.

Emmons, R.A. and McCullough, M.E. (2003) *Counting Blessings Versus Burdens: An Experimental Investigation of Gratitude and Subjective Well-Being in Daily Life*. US National Library of Medicine National Institutes of Health.

Ernst and Young (2014) *Business Pulse: Exploring the Dual Perspectives of The Top Ten Risks and Opportunities in 2013 and Beyond*. EY Global.

Estrada, C.A., Isen, A.M. and Young, M.J. (1994). Positive Affect Improves Creative Problem Solving and Influences Reported Source of Practice Satisfaction in Physicians. *American Psychological Association*

Estrada, C.A., Isen, A.M. and Young, M.J. (1997). *Organizational Behavior and Human Decision Processes*, 72(1).

Fenton, J. (1990) *101 Ways to Boost Your Business Performance*. London: Mandarin.

Ferguson, K.I. (2015) *A Study of Safety Leadership and Safety Governance for Board Members and Senior Executives*. PhD thesis, Queensland University of Technology. eprints.qut.edu.au/81349/.

Festinger, L. (1955) Social Psychology and Group Processes. *Annual Review of Psychology*, 6, 187–216.

Festinger, L. (1957) *A Theory of Cognitive Dissonance*. Stanford, CA: Stanford University Press.

Festinger, L. and Carlsmith, J.M. (1959). Cognitive Consequences of Forced Compliance. *Journal of Abnormal and Social Psychology*.

Fiedler, F.E. (1964) A Contingency Model of Leadership Effectiveness. In L. Berkowitz (ed.) *Advances in Experimental Social Psychology*. New York: Academic Press.

Fiedler, F.E. (1967) *A Theory of Leadership Effectiveness*. New York: McGraw-Hill.

Fletcher, J. (2004) The Paradox of Postheroic Leadership. *Leadership Quarterly*, 15, 647–661.

Flin, R., Mearns, K., O'Connor, P., and Bryden, R. (2000) Measuring Safety Climate. Identifying the Common Features. *Safety Science*, 34(1–3), 177–193.

Frank, R.H. (2008) *The Economic Naturalist: Why Economics Explains Almost Everything*. London: Virgin Books.

Friedman, M. (1970) The Social Responsibility of Business is to Increase Its Profits. *New York Times Magazine*, September 13.

Frigo, M.L. and Anderson, R.J. (2011) Strategic Risk Management: A Foundation for Improving Enterprise Risk Management and Governance. *Journal of Corporate Accounting & Finance*. DOI: 10.1002/jcaf.20677.

Fukuyama, F. (1996) *Trust: The Social Virtues and Creation of Prosperity*. New York: Free Press.

Furedi, F. (1997) *The Culture of Fear; Risk Taking and the Morality of Low Expectations*. London: Cassell.

Furedi, F. (2002) *The Culture of Fear*. London: Continuum.

Gadd, S. (2002) *Safety Culture: A Review of the Literature*. HSL/2002/25. hse.gov.uk/research/hsl_pdf/2002/hsl02-25.

Gardner, D. (2008) *Risk: The Science and Politics of Fear*. London: Virgin Books.

Geller, S. (2001) *The Psychology of Safety Handbook*. London: Lewis Publishers.

George, B. (2003) *Authentic Leadership: Rediscovering the Secrets to Creating Lasting Value*. San Francisco, CA: Jossey-Bass.

Gerstner, L (2013) *Who Says Elephants Can't Dance?* New York: Harper-Collins

Giddens, A. (1991) *Modernity and Self-Identity*. Cambridge, UK: Polity Press.

Gilbert, D. (2007) *Stumbling on Happiness*. New York: Vintage.

Gilley, J., Boughton, N.W. and Maycunich, A. (1999) *The Performance Challenge*. New York: Basic Books.

Gladwell, M. (2002) *The Tipping Point: How Little Things Can Make a Big Difference*. Boston, MA: Little Brown.

Gladwell, M. (2005) *The Tipping Point*. London: Penguin.

Glendon, A.I. and Stanton, N.A. (2000) Perspectives on Safety Culture. *Safety Science*, 34, 193–214.

Goldstein, N.J., Martin, S.J. and Cialdini, R.B. (2007) *Yes! The Power of Persuasion*. London: Profile.

Goleman, D. (1995) *Emotional Intelligence*. London: Bantam Books

Grawitch, M.J., Munz, D.C., Elliott, E.K. and Mathis, A. (2003). Promoting Creativity in Temporary Problem-Solving Groups: The Effects of Positive Mood and Autonomy in Problem Definition on Idea-Generating Performance. *American Psychological Association*.

Grint, Keith (2000) *The Arts of Leadership*. Oxford, UK: Oxford University Press.

Grint, Keith (2005) *Leadership: Limits and Possibilities*. Basingstoke, UK: Palgrave Macmillan.

Grint, K. (2007) Learning to Lead: Can Aristotle Help Us to Find the Road to Wisdom? *Leadership*, 3(2), 231–46.

Grosz, S. (2013) *The Examined Life*. London: Chatto & Windus.

Hackman, J.R. and Oldham, G.R. (1976) Motivation Through the Design of Work: Test of a Theory. *Organizational Behavior and Human Performance*, 16(2).

Hackman, J.R. and Oldham, G.R. (1980) *Work Redesign*. Reading, MA: Addison-Wesley.

Hale, A.R. (2000) Culture's Confusions. *Safety Science*, 34, 1–14.

Hanh, T.H. (1991) *The Miracle of Mindfulness*. London: Random House.

Harrington, H.J. (1991) *Business Process Improvement: The Breakthrough Strategy for Total Quality, Productivity, and Competitiveness*. New York: McGraw-Hill Education.

Harter, J.K., Schmidt, F.L., Killham, E.A., and Agrawal, S. (2009). *Q12 Meta-Analysis: The Relationship Between Engagement at Work and Organisational Outcomes*. Gallup.

Hasson, F. (2012) *Delegating and Supervising Unregistered Professionals: The Student Nurse Experience*. ResearchGate. Nurse Education Today.

Hasson, U., Ghazanfar, A.A., Galantucci, B., Garrod, S. and Keysers, C. (2012) *Brain-to-brain Coupling: A Mechanism for Creating and Sharing a Social World.* Neuroscience Institute, Princeton University.

Health & Safety Executive (1995) *Improving Compliance with Safety Procedures: Reducing Industrial Violations.* hse.gov.uk/research.

Health & Safety Executive (2002) *Young People's Attitudes to Health and Safety at Work.* hse.gov.uk/research.

Health & Safety Executive (2009) *Reducing Error and Influencing Behaviour* HSG48. hse.gov.uk/research.

Health & Safety Executive (2010) *AALS Inspector Guidance Note – IGN 1.08.* hse.gov.uk/research.

Health and Safety Laboratory, Human Factors Group (2002) *Safety Culture: A Review of the Literature.* HSL/2002/25.

Heemstra, F. and Kusters, R. (1996) Dealing with Risk: A Practical Approach. *Journal of Information Technology*, 11, 333–346.

Heifetz, R. and Laurie, D. (1997) *The Work of Leadership.* Boston, MA: Harvard Business Review Press.

Heinrich, H.W. (1931) *Industrial Accident Prevention: A Scientific Approach.* New York: McGraw-Hill.

Heller, E. (2000) *An Agapas Stravōnesai! Diadromḗ stē Sýnchronē Logotechnía.* Empeipia.

Herrero, L. (2008) *Viral Change.* London: Meeting Minds.

Hersey, P. and Blanchard, K.H. (1988) *Management and Organizational Behavior: Utilizing Human Resources.* Englewood Cliffs, NJ: Prentice-Hall.

Hofstede, G. (1980) *Culture's Consequences: International Differences in Work-Related Values.* Thousand Oaks, CA: Sage.

Hofstede, G. (2002) *Exploring Culture: Exercises, Stories and Synthetic Cultures.* London: Nicholas Brealey.

Hollingworth, P. (2016) *The Light and Fast Organisation.* Milton, UK: Wiley.

Hollnagel, E. (2014) *Safety-I and Safety-II.* Farnham, UK: Ashgate.

Hopkins, A. (2005) *Safety, Culture and Risk.* Sydney: CCH.

Hopkins, A. (2008) *Failure to learn: the BP Texas City Refinery Disaster.* North Ryde: CCH.

Hopkins, A. (2012) *Disastrous Decisions.* North Ryde: CCH.

Huff, D. (1954) *How to Lie with Statistics.* New York: W.W. Norton.

Hunt, R. (2013) *Be Silent or Be Killed: The True Story of a Scottish Banker Under Siege in Mumbai's 9/11.* Edinburgh, UK: Corskie Press.

Huppert, F A. (2008) *Psychological Well-being: Evidence Regarding its Causes and Consequences.* Cambridge, UK: Cambridge University Press.

Huseman, R.C., Hatfield, J.D. and Miles, E.W. (1987). A New Perspective on Equity Theory: The Equity Sensitivity Construct. *Academy of Management Review*, 12(2), 222–234.

Ilic, D. (2013) Listening to Favorite Music Improves Endothelial Function in CAD. *European Society of Cardiology.* September.

Institute of Directors & Health and Safety Commission (2007) *Leading Health and Safety at Work: Leadership Actions for Directors and Board Members.* HSE, UK.

Jochelson K, 2007. *Paying the Patient: Improving Health Using Financial Incentives.* The Kings Fund.

Jones, A. (2005) The Anthropology of Leadership: Culture and Corporate Leadership in the American South. *Leadership*, 1(3), 259–278.

Jones, A. (2006) Developing What? An Anthropological Look at the Leadership Development Process. *Leadership*, 2(4), 481–498.

Jones, M.R. (1953) *Current Theory and Research in Motivation*. Lincoln, NE: University of Nebraska Press.

Kahneman, D. (2011) *Thinking, Fast and Slow*. London: Penguin.

Kahneman, D. and Tversky, A. (2000) *Choices, Values and Frames*. London: Cambridge University Press.

Kant, I. (1993 [1785]) *Grounding for the Metaphysics of Morals*, translated by J.W. Ellington. Indianapolis, IN: Hackett Publishing.

Kardashian, K. (2013) *When Retailers Do Good, Are Consumers More Loyal?* Tuck School of Business. December. tuck.dartmouth.edu/news/articles/when-retailers-do-good-are-consumers-more-loyal.

Kegan, R. and Laskow-Lahey, L. (2009) *Immunity to Change*. Boston, MA: Harvard Business Review Press.

Kirkpatrick, S.A. and Locke, E.A. (1991) *Leadership: Do Traits Matter?* Academy of Management Executive.

Kiron, D. (2015) *Sustainability & Innovation Global Executive Study*. MIT Sloan Management Review.

Klein, G. (2003) *The Power of Intuition*. New York: Doubleday.

Klein, G. (2011) *Streetlights and Shadows*. Cambridge, MA: MIT Press.

Komaki, J.L. (1998) *Leadership from an Operant Perspective*. London: Routledge.

Komaki, J.L., Collins, R.L. and Penn, P. (1982) The Role of Performance Antecedents and Consequences in Work Motivation. *Journal of Applied Psychology*, 67(3), 334–340.

Kortum E. (2012) A Need to Broaden Our Perspective to Address Workers' Health Effectively in the 21st Century. *Journal of Industrial Health*, 50, 71–72.

Kotter, J. (1996) *Leading Change*. Boston, MA: Harvard Business Review Press.

Kouzes, J. and Pozner, B. (2008) *The Leadership Challenge*. San Francisco, CA: Jossey Bass.

KPMG and IDH (2014) *A New Vision of Value: Connecting Corporate and Societal Value Creation*. KPMG Global.

LaClair, J. and Rao, P. (2002) *Helping Employees Embrace Change*. McKinsey Quarterly. Galegroup.

Lancaster, T., Stead, L., Silagy, C. and Sowden, A. (2000) Effectiveness of Interventions to Help People Stop Smoking. *British Medical Journal*, 321, 7257, 355–358.

Levering, R. (1988) *A Great Place to Work*. New York: Random House.

Lewin, K. and Cartwright, D. (ed.) (1951) *Field Theory in Social Science*. New York: Harper.

Lewin, K. and Lewin, G.W. (eds) (1948) *Resolving Social Conflicts: Selected Papers on Group Dynamics [1935–1946]*. New York: Harper and Brothers.

Lockwood, N.R. (2007) Leveraging Employee Engagement for Competitive Advantage. *SHRM Research Quarterly*, 2–11.

Löfstedt, R. (2011) *Reclaiming Health and Safety for All: An Independent Review of Health and Safety Legislation*. Health and safety reform. Gov.com.

Long, R. (2012) *For the Love of Zero: Human Fallibility and Risk*. Australia: Scotoma Press.

Long, R. (2012) *Risk Makes Sense*. Australia: Scotoma Press.

Lupton, D. (1999) *Risk*. Abingdon, UK: Routledge.

Lykken, D. and Tellegen, A. (1996) Happiness is a Stochastic Phenomenon. *Psychological Science*, 7(3), 186–189.

Lyubomirsky, S. (2005) *Pursuing Happiness: The Architecture of Sustainable Change*. Educational Publishing Foundation.

Lyubomirsky, S., King, L.A. and Diener, E. (2005). The Benefits of Frequent Positive Affect: Does Happiness Lead to Success? *Psychological Bulletin*, 131, 803–855.

Maccoby, M. (2004) Narcissistic Leaders: The Incredible Pros, The Inevitable Cons. *Harvard Business Review*, 82(1) January–February, 92–101.

Machiavelli, N. (2003). *The Prince*. London: Penguin Classics.

Mack, A. and Rock, I. (1992) *Inattentional Blindness*. Cambridge, MA: MIT Press.

Macpherson, C. (2017) *The Change Catalyst*. London: Wiley.

Mansley, M. (2002) *Health and Safety Indicators for Institutional Investors*. Report to the HSE, Claros Consulting.

Marciano, P. (2010) *Carrots and Sticks Don't Work: Build a Culture of Employee Engagement with the Principles of RESPECT*. New York: McGraw-Hill Education.

Marsh, T. (2013) *Talking Safety*. Gower, UK: Farnham.

Marsh, T. (2014) *Total Safety Culture*. Manchester, UK: RyderMarsh.

Maxmin, J. and Zuboff, S. (2002) *The Support Economy*. London: Penguin.

McCall, M.W. and Lombardo, M.M. (eds) (1978) *Leadership: Where Else Can We Go?* (pp. 87–99). Durham, NC: Duke University Press.

McCormick, J. and Tiffin, J. (1965) *Industrial Psychology*. Englewood Cliffs, CA: Prentice-Hall.

McGregor, D. (1960) *The Human Side of Enterprise*. New York: McGraw-Hill Education.

McLaughlin, C. and Davidson, G. (1994) *Spiritual Politics: Changing the World from the Inside Out*. New York: Ballantine Books.

Medibank Private (2005) The Health of Australia's Workforce, at www.trenchhealth.com.au/articles/MEDI_Workplace_Web_Sp.pdf.

Meindl, J., Ehrlich, S. and Dukerich, J. (1985) The Romanticism of Leadership. *Administrative Science Quarterly*, 30(1).

Morrison, R.L. and Macky, K. (2017) *The Demands and Resources Arising from Shared Office Spaces*. ScienceDirect.

Mulholland R.E., Sheel, A.G. and Groat, S. (2005) *Research Report 306*. Sudbury, UK: HSE Books.

Myers, D.G. (2005). Comment on Goodbye Justice, Hello Happiness: Welcoming Positive Psychology to the Law. *Deakin Law Review* 27.

Myers, D.G. and Diener, E. (2005) *Who is Happy?* Champaign, IL: University of Illinois Press.

Nahrgang, J., Morgeson F. and Hofman, D. (2011) Safety at Work: A Meta-analytic Investigation of the Link Between Job Demands, Job Resources, Burnout, Engagement, and Safety Outcomes. *Journal of Applied Psychology*, DOI: 10.1037/a0021484.

Neilson, G.L. and Pasternack, B.A. (2005) *Keep What's Good, Fix What's Wrong, and Unlock Great Performance*. New York: Crown Business.

Norman, G.R. (1985) *Assessing Clinical Competence*. New York: Springer.

O'Neill, O. (1996). *Towards Justice and Virtue*. Cambridge, UK: Cambridge University Press.

Pearson, D. (2001). Motivating Change. *Nursing Standard*, 16(2).

Peters, T. and Waterman, R. (1982) *In Search of Excellence*. New York: HarperCollins.

Pink, D. (2011) *Drive*. Edinburgh, UK: Canongate Books.

Pondy, L.R. (1978) Leadership is a Language Game. In McCall, M.W. and Lombardo, M.M. (eds), *Leadership: Where Else Can We Go?* (pp.87–99). Durham, NC: Duke University Press.

Porter, M.E. (2011) Creating Shared Value: Redefining Capitalism and the Role of the Corporation in Society. *Harvard Business Review*. FSG CSV Leadership Summit, June.

Pressfield, S. (2002) *The War of Art*. New York: Black Irish.

Quilley, A. (2006) *The Emperor has no Hard Hat*. Sherwood Park: Safety Results.

Quilley, A. (2010) *Creating and Maintaining a Practical-Based Safety Culture*. Sherwood Park: Safety Results.

Quoidbach, J., Gilbert, D.T. and Wilson, T.D. (2013) The End of History Illusion. *Science*, 339, 96–98.

Reason, J. (1997) *Managing the Risks of Organizational Accidents*. Farnham, UK: Ashgate.

Reason, J. (1998) *Achieving a Safe Culture: Theory and Practice: Work and Stress*. Farnham, UK: Ashgate.

Reason, J. (2008) *The Human Contribution*. Farnham, UK: Ashgate.

Reber, A.S., Reber, E. and Allen, R. (2009) *Penguin Dictionary of Psychology*. 4th Edition. London: Penguin.

Reddy, S. (2013) *The Rana Plaza Building Collapse . . . 100 Days On*. ILO Global. ilo.org/global/.

Reid, M. (2008) Behind the 'Glasgow Effect'. *Bulletin of the World Health Organization*, 89(10). DOI: 10.2471/BLT.11.021011.

Robinson, D., Bevan, S. and Barber, L. (1997) *Keeping the Best. A Practical Guide to Retaining Key Employees*. IES.

Ronald, L. (1999) *Identifying the Elements of Successful Safety Programs: A Literature Review*. Workers' Compensation Board of British Columbia.

Ropeik, D. (2010) *How Risky Is it, Really? Why Our Fears Don't Always Match the Facts*. New York: McGraw-Hill.

Ropeik, D. (2002) Understanding Factors of Risk Perception, Nieman Reports, Vol 56.

Ropeik D. Understanding Factors of Risk Perception, Nieman Reports (Winter 2002). Harvard business review

Roughton, J. and Mercurio, J. (2001) *In Developing an Effective Safety Culture*. Oxford, UK: Butterworth-Heinemann.

Rowland, D. and Higgs, M. (2008) *Sustaining Change*. Chichester, UK: John Wiley.

Schein, E.H. (1985) *Organizational Culture and Leadership*. San Francisco, CA: Jossey-Bass.

Schein, E.H. (2002) *The Anxiety of Learning*. Boston, MA: Harvard Business Review.

Schein, E.H. (2012) *The Corporate Culture Survival Guide*. San Francisco, CA: Jossey-Bass.

Schein, E.H. (2013) *Humble Inquiry*. Oakland, CA: Berrett-Koehler.

Schulte, P. and Vainio, H. (2010) Well-being at Work – Overview and Perspective. *Scandinavian Journal of Work, Environment & Health*, 36. DOI:10.5271/sjweh.3076.

Schwartz, T. (2010) *The Way We're Working Isn't Working*. London: Simon & Schuster.

Sharman, A. (2016) *From Accidents to Zero: A Practical Guide to Improving Your Workplace Health and Safety Culture*. Abingdon, UK: Routledge.

Sharman, A. (2017) *Mind Your Own Business: What Your MBA Should Have Taught You About Workplace Health and Safety*. Edinburgh, UK: Maverick Eagle Press.

Sharman, A. and Marsh, T. (2016) *Safety Savvy*. Edinburgh, UK: Maverick Eagle Press.

Shewhart, W.A. (1931) *Economic Control of Quality of Manufactured Product*. Martino Fine Books.

Skinner, B.F. (1953) *Science and Human Behaviour*. New York: The Free Press.

Slovic, P. (2000) *The Perception of Risk*. London: Earthscan.

Slovic, P. (2010) *The Feeling of Risk: New Perspectives on Risk Perception*. London: Earthscan.

Smith Institute (2005) Seminar Series 2: Healthy Living, Health in the Workplace. The Oxford Health Alliance, www.oxha.org/meetings/meetings-attended-by-oxha/smith-institute-seminar-series-2013-2018your-good-health2019.

Smith Institute (2005) *Public Sector Procurement and the Public Interest*. RBC Capital Markets – Business & Economics.

Sparr, J.L. and Sonnentag, S. (2008) Fairness Perceptions of Supervisor Feedback, LMX, and Employee Well-Being at Work. *European Journal of Work and Organizational Psychology*, 17(2), 198–225.

Spector, P. (1997). *Job Satisfaction: Application, Assessment, Causes and Consequences*. Thousand Oaks, CA: Sage.

Spector, P.E. (2002) *Employee Control and Occupational Stress*. Thousand Oaks, CA: Sage.

SSEE (2014) *From the Stockholder to the Stakeholder: How Sustainability Can Drive Financial Outperformance*. University of Oxford/Arabesque.

Stogdill, R.M. (1948) Personal Factors Associated with Leadership. *Journal of Psychology*, 25, 35–71.

Stogdill, R.M. (1974) *Handbook of Leadership: A Survey of Theory and Research*. New York: Free Press.

Szalai, A. and Andrews, F.M. (1980) *The Quality of Life: Comparative Studies*. London: Sage. trove.nla.gov.au/work/10038628?q&versionId=11661889.

Taleb, N. (2007) *The Black Swan: The Impact of the Highly Improbable*. London: Random House.

Taylor, J.B. (2010) *Safety Culture: Assessing and Changing the Behaviour of Organisations*. Farnham, UK: Gower.

Thaler, R.H. and Sunstein, C.R. (2009) *Nudge*. London: Penguin.

Thompson, D. (2005) *Restoring Responsibility*. Cambridge, UK: Cambridge University Press.

Tiffin, J. and McCormick, E. (1971) *Attitude and Motivation*. London: HMSO.

Tilman, L. (2012). *Risk Intelligence: A Bedrock of Dynamism and Lasting Value Creation*. European Financial Review. Deloitte.

Topchik, G.S. (2001) *Managing Workplace Negativity*. AMACOM.

Townsend, A.S. (2013) *Safety Can't Be Measured*. Farnham, UK: Gower.

Turner, B.A. and Pidgeon, N. (1997) *Man-Made Disasters*. Oxford, UK: Butterworth-Heinemann.

Tzu, Sun (2009) *The Art of War*. London: Penguin Classics

Vaughan-Jones, H. and Barham, L. (2009) *Healthy Work*. London: BUPA

Waddell, G. and Burton, A. K. (2006) *Is Work Good for Your Health and Wellbeing?* London: The Stationery Office.

Weick, K. (1979) *The Social Psychology of Organizing*. New York: McGraw Hill.

Weiss, H.M. and Cropanzano, R. (1996). Affective Events Theory: A Theoretical Discussion of the Structure, Causes and Consequences of Affective Experiences at Work. January.

Wheatley, M.J. and Kellner-Rogers, M. (1996) *A Simpler Way*. San Francisco, CA: Berret-Koehler.

WHO (1999) *The Impact of Globalisation on Health and Safety at Work*. Geneva: World Health Organization.

Wiseman, R. (2009) *59 Seconds: Think a Little, Change a Lot*. London: Macmillan.

Wood, M. (2005) The Fallacy of Misplaced Leadership. *Journal of Management Studies*, 42(6), 1101–1121.

Wood, M. and Ladkin, D. (2008) The Event's the Thing: Brief Encounters with the Leaderful Moment. In K. Turnbull James and J. Collins (eds) *Leadership Perspectives: Knowledge into Action*. Houndmills, UK: Palgrave Macmillan.

Yoshie, M. et al., (2009) Exp Bron Res, 34, 23–29.

Yukl, G. (2002) *Leadership in Organizations*. 5th Ed. Upper Saddle River, NJ: Prentice-Hall.

Zigenfus, G.C., Yin, J. and Giang, G.M. (2001) Effectiveness of early physical therapy in the treatment of acute low back musculoskeletal disorders. *Journal of Occupational & Environmental Medicine*, 42.

Index

Page numbers in *italics* refer to figures. Page numbers in **bold** refer to tables.